IMAGES of America
THREE MILE ISLAND

General Public Utilities Nuclear Corporation and Three Mile Island spokesman Gordon Tomb takes some time away from the job to indulge in a bit of fly-fishing on the Susquehanna River, upstream of the nuclear power plant, in this image taken on August 7, 1987. Tomb currently serves as a senior fellow for the Commonwealth Foundation in Harrisburg, Pennsylvania. (Courtesy of the Historical Society of Dauphin County.)

ON THE COVER: This is a nighttime view of Three Mile Island with the cooling towers of the Unit 1 reactor in the foreground. This particular view was taken on January 2, 1980, while both the Unit 1 and Unit 2 reactors were dormant following the nuclear accident that had occurred on March 28, 1979. (Courtesy of the Historical Society of Dauphin County.)

Images of America

Three Mile Island

Erik V. Fasick

Copyright © 2018 by Erik V. Fasick
ISBN 978-1-4671-0285-8

Published by Arcadia Publishing
Charleston, South Carolina

Library of Congress Control Number: 2018951517

For all general information, please contact Arcadia Publishing:
Telephone 843-853-2070
Fax 843-853-0044
E-mail sales@arcadiapublishing.com
For customer service and orders:
Toll-Free 1-888-313-2665

Visit us on the Internet at www.arcadiapublishing.com

To Danielle, Jack, and Julia. This book would not have been possible without your endless patience and understanding.

Contents

Acknowledgments 6

Introduction 7

1. In the Beginning 9
2. The Accident 35
3. Fallout 77

ACKNOWLEDGMENTS

As always, I am greatly indebted to the Board of Trustees of the Historical Society of Dauphin County, and the society's staff, for providing me with access to their vast collection of resources. As such, all author royalties from this volume will be donated to the Historical Society of Dauphin County. These images are the result of the extraordinary resolve and talent of the Allied Pix photographers who remained steady and diligent, often in the face of chaos and uncertainty. However, the context for these images was provided, first and foremost, by the investigative work of Mary Bradley. I owe a huge debt of gratitude to her work reporting for the *Patriot News*, the result of which is an extensive collection of Three Mile Island materials housed at the Historical Society of Dauphin County. Unless otherwise noted, all images in this volume appear courtesy of the Historical Society of Dauphin County.

In addition, this work was given depth by the professional reporting of the staff of the *Patriot News*, the Associated Press, United Press International, the *Carlisle Sentinel*, the *Allentown Morning Call*, the *Lebanon Daily News*, the *Philadelphia Daily News*, the *Philadelphia Inquirer*, the *Washington Post*, and the *Pittsburgh Post-Gazette*. I would also like to thank the Nuclear Regulatory Commission for its copious storehouse of information related to Three Mile Island and the nuclear industry, which they have made readily available to the public. Also, I need to extend a thank-you to Three Mile Island Alert for the resources that its has made available through its website, along with the writings of Scott Johnson, which helped me to make sense of the events of March 28, 1979. I would also be remiss if I did not thank Geary Huntsberger, Roger Thomas, and Ronald Kopp for their recollections and anecdotes. Lastly, I would like to thank my title manager Caitrin Cunningham and everyone at Arcadia Publishing, all of whom have shown an extraordinary amount of professionalism and patience in guiding me through this project. Again, and most sincerely, thank you all.

INTRODUCTION

The Three Mile Island Nuclear Power Plant is comprised of two separate nuclear reactors. The Unit 1 reactor began commercially producing electricity in September 1974. The timing was fortunate for the nuclear industry, as it could point to the power plant as a success during a time when the United States had recently emerged from a protracted fight with OPEC and the oil embargo that had led to long lines at the gas pumps and the cost of crude oil skyrocketing. However, what should have been promising times for the owners of Three Mile Island had turned to hand-wringing, as both units of the power plant had gone grossly over budget. The completion of the Unit 1 power plant, which was initially estimated to cost $110 million, would ultimately carry a price tag of over $400 million by the time it began producing electricity in late 1974. Construction of the Unit 2 power plant began in late 1969, but it would not begin producing commercial electricity until December 30, 1978. The total cost of the power plant would balloon to $1 billion. Soon after the Unit 1 reactor came on line and began producing electricity, Walter Creitz, president of Metropolitan Edison, was asked by an unidentified reporter on a press pool tour about the long delays in bringing the Unit 2 reactor on line; he simply replied, "Because we don't have the money." This quote is from the November 14, 1974, *Lebanon Daily News*.

The ownership, and the financial responsibility for Three Mile Island, belonged in part to Metropolitan Edison, which owned 50 percent of the nuclear power plant. The remaining 50 percent was evenly divided between Pennsylvania Electric Company and Jersey Central Power and Light Company. All three entities served as subsidies of General Public Utilities Corporation.

The promise of a highly profitable business venture shattered just before 4:00 a.m. on March 28, 1979, as a water pump failed, a valve failed to close, and human errors led to a loss of coolant in the reactor core, which resulted in the partial meltdown of the Unit 2 reactor. The errors could have been avoided, as similar problems arose in the testing of similar reactors designed by Babcock and Wilcox. However, the design engineers believed that the plant was incapable of failing. The power plant was designed to run on its own, and little effort was put into responding to emergencies that they felt would never arise. Control room operators were taught to handle "situation normal" scenarios, while managing emergency situations was left out of the training. As a result, it took eight hours to properly diagnose and stabilize the reactor core after things started to go awry.

Although a general emergency had been enacted, and the plant was placed on lockdown, Metropolitan Edison attempted to downplay the incident. "It's nothing we can't take care of," stated vice president Jack Herbein. Over the next few days, information released from Metropolitan Edison to the media became untrustworthy as reports of radiation releases were contradicted by reports from the Nuclear Regulatory Commission. On the morning of March 30, an unannounced release of radioactive gas, resulting in a plume several miles long, was detected as it moved toward population centers to the north and east. Again, conflicting information on the extent of the radiation release prompted Gov. Dick Thornburgh to issue an evacuation for pregnant women and young children within a five-mile radius of the plant. Schools located within a 10-mile radius began to voluntarily dismiss students. Shelters were established to handle the evacuees. Thousands more grabbed what they could and left the area. The Nuclear Regulatory Commission appointed Harold Denton to provide a singular calming voice to provide the media—and, ultimately, the public—with an assessment of the situation that could be readily understood.

On March 16, 1979, just 12 days prior to the accident, *The China Syndrome* was released in theaters. The film details the near-miss of a meltdown at a fictional nuclear power plant in California, along with a cover-up of cost-cutting measures taken during the construction of the plant. In a moment of irony, a scene in the film shows characters discussing the possibility of a meltdown of the reactor and delivering a line saying that the resulting disaster would "render an area the size of Pennsylvania uninhabitable." I vividly recall, as a seven-year-old child, sitting in the movie theater with my parents the week following the accident. And although everyone in the theater knew that the line was coming, there was still a collective gasp from the audience when it was delivered. Fictional or not, the fear was real.

Sensing a need for calm, the decision was made for Pres. Jimmy Carter to travel to Three Mile Island on April 1 to tour the plant and obtain an on-the-ground assessment of the situation. Carter had some prior experience as a naval officer who was present during the cleanup of the partial meltdown of the NRX reactor at Chalk River in Canada. While the immediate threat of a meltdown had passed, and radiation levels had seemingly subsided, a new fear had emerged. A hydrogen bubble was discovered in the reactor vessel; under the right conditions, this bubble could explode, breaching the reactor and containment building and issuing radioactive material into the environment. It was a calculated gamble to send the president of the United States into a potentially dangerous situation, but the administration felt that the risk was minimal and that putting on a brave face would allay the public's fears. Although Carter's tour of the plant was brief, it showed members of the public that they would survive Three Mile Island.

As President Carter left, so did the hydrogen bubble. The reactor began a long cooling phase that would take weeks to enter into a cold shutdown. As things cooled within the power plant, public opposition to nuclear power began to heat up. What had previously been small localized efforts to denounce nuclear energy had suddenly blossomed into a national rallying cry against the nuclear industry. Three Mile Island symbolized the dangerous side of nuclear power. The anti-nuclear rallies that once attracted dozens had grown to include hundreds, then thousands, as celebrities such as Jane Fonda lent their voices to the cause. As the crowds grew, politicians seeking public backing for their candidacies tailored their messages to ride the anti-nuclear wave. Three Mile Island, and the ominous cooling towers, provided a backdrop and a talking point for Walter Mondale, Jesse Jackson, Jerry Brown, and John Anderson as the nuclear power plant became a stop on the presidential campaign trail.

As the Unit 2 reactor lay damaged beyond repair, the focus shifted to the Unit 1 reactor, which had been out of commission in March 1979 for refueling. The Unit 1 reactor was not brought back into service immediately after the accident but was left in legal limbo pending a review to determine if it, too, was a safety hazard. The protracted battle to restart the Unit 1 reactor would play out over six years before a ruling was made by the Supreme Court in October 1985 to allow the Unit 1 reactor to once again produce electricity. While this battle was being waged in the courts, the cleanup of the Unit 2 reactor continued at a slow pace that would last for more than 14 years at a total cost of $1 billion.

One

IN THE BEGINNING

On November 14, 1974, Walter Creitz, president of Metropolitan Edison, was giving a tour of the Three Mile Island Nuclear Power plant to a group of journalists. Creitz made a bold prediction to the visiting group, exclaiming that "nuclear power will revolutionize the world." Nuclear power would revolutionize the world, and Three Mile Island would be ground zero for that revolution.

In this aerial view of the Susquehanna River, Three Mile Island is visible in its entirety with the nuclear power plant at its northern end. The large island to the west is Shelley's Island, which is largely farmland except for the summer cottages that line its outer edges. The northern access to Three Mile Island is reached by a bridge spanning to Londonderry Township, located in the southernmost portion of Dauphin County. This western edge of the township is dominated by farmland with a small sprinkling of houses along Route 441. This route travels south along the river, past the southern access bridge to Three Mile Island, and toward the small community of Falmouth (upper left), just over the Lancaster County line. A portion of Goldsboro (upper right) is visible on the western shore of the Susquehanna River.

Three Mile Island, along with Shelley, Hill, and several smaller islands that surround the three larger islands, are all within the boundaries of Londonderry Township in Dauphin County. Three Mile Island, named for its approximate distance to Middletown Borough, was utilized as farmland and produced choice tobacco crops for Col. James Duffy during the late 19th century. As a result, the island was called "Duffy's Island" for many years and even held on to the moniker after the colonel's death in 1888. Prior to this, the island was labeled Conewago Island in the 1875 atlas of Dauphin County; it was named for the creek that empties into the Susquehanna River near the island's southern end.

In 1905, the York Haven Water and Power Company purchased the island, after it had passed from the estate of Col. James Duffy, with the intent of building a dam from the island to the eastern shore. In 1926, the Metropolitan Edison Company absorbed the York Haven Water and Power Company and acquired Three Mile Island in the process. However, the power companies were not interested in the island itself but rather its location within the Susquehanna River and the possibilities it held in relation to their power plants. As a result, the island was leased to farmers who made their residences on the island and even constructed a one-room schoolhouse.

Londonderry Township dates to 1767, when it was split from Derry Township, which was then part of Lancaster County. The township was incorporated into Dauphin County when it was established in 1785. The southernmost township in Dauphin County, Londonderry Township has existed largely as a rural collection of small farms for the last 200 years. Many of the farmers are of Pennsylvania German descent, raising dairy cattle and rotated crops such as corn and soybeans. The small villages of Rocktown and Gainesburg could once be found in the township along the East Harrisburg Pike, but they have since faded away. The Pennsylvania Canal also ran through the township along the bank of the Susquehanna River. The population of the township has steadily remained at around 5,000 inhabitants since the nuclear accident in 1979.

Dairy farmer Ronald Kopp stands in his fields near Colebrook Road in Londonderry Township in this image taken in March 1989. Kopp is a second-generation dairy farmer on a small family farm that was purchased by his father in 1946. The Kopp farm is typical of the non-commercialized farms that could be found in Londonderry Township during the last half of the 20th century.

It is tee time at the Sunset Golf Course in Londonderry Township, with the cooling towers of the Three Mile Island nuclear power plant visible in the background. The golf course, in part, dates to the 1930s as a nine-hole course laid out for the officers of the Middletown Air Depot. The grounds were acquired in 1968 by Londonderry Township for use as a public golf course with adjacent recreational facilities.

Goldsboro, which was a community of 600 residents during the 1970s, is situated along the western shore of the Susquehanna River opposite Three Mile Island, whose cooling towers seemingly loom over the tiny borough and visually dominate the landscape. Founded in 1850 by Major Goldsboro, an engineer who worked on the Pennsylvania Railroad, the relatively isolated community became a destination for those seeking leisurely activities on the Susquehanna River and its many islands. Prior to the nuclear accident, Goldsboro's only other entries into the national spotlight came from a 47-round prizefight in 1867 between pugilists Sam Collyer of Baltimore and John McGlade of New York City and as the hometown of former major-league baseball player Greg Gross.

Above, Geary Huntsberger, a farmer from nearby Etters, pilots his barge away from the dock in Goldsboro and heads upstream to nearby Hill Island, which is situated in the Susquehanna River one mile northeast of Three Mile Island. Huntsberger, shown below in the freshly plowed fields on Hill Island, leased 170 acres of land on Three Mile Island, where he grew corn and wheat, from Metropolitan Edison during the 1960s. In 1967, when Three Mile Island was selected as the site for a nuclear power plant, Metropolitan Edison discontinued his lease for the farmland but contracted him out for the removal of cabins and trailers of individuals who leased space on the island for their summer "cottages." Huntsberger ferried the cabins and trailers on his barge to nearby Shelley Island, also owned by Metropolitan Edison.

Construction of the Unit 1 nuclear power plant on Three Mile Island began in May 1968 with an estimated price tag of $100 million, with hopes that it would be completed by March 1971. However, it would not be until September 1974 that commercial electricity was produced at the plant at a completed cost of over $400 million. The permit granting approval for construction of the Unit 2 nuclear power plant was issued in November 1969, but the Unit 2 nuclear power

plant did not begin producing commercial electricity until December 30, 1978. The total cost of the entire power plant exceeded $1 billion. In this image, the component buildings of Unit 1 and its cooling towers are located on the northern end of the island, to the left, with the adjoining Unit 2 to the right.

These two images show the Unit 2 containment building while it was under construction in October 1974. When completed, the Unit 2 containment building reached a height of 200 feet with an outside diameter of 137 feet. The walls of the structure are 3½ feet thick and constructed of steel-reinforced concrete. The strength of the containment building walls was touted as being able to withstand the impact of a Boeing 727 jet aircraft crashing into them. This assertion has never been tested, even with Harrisburg International Airport in proximity. The size and security of the building was necessary, as the containment building housed the nuclear reactor and uranium fuel used to initiate the production of electricity for the power plant.

This photograph was taken on October 22, 1974, and shows an interior view of one of the two cooling towers for the Unit 2 reactor before the reactor and towers had become operational. Built at a cost of $7 million per tower and constructed out of reinforced concrete, each tower reaches a height of 372 feet—as tall as a 37-story building. Each tower has a diameter of 380 feet at the base and 180 feet at the top, giving the overall structure its iconic hyperbolic shape. The base of each functioning tower has a black ring structure, which consists of fin-like baffles or louvers. The cooling towers provide an ecological solution to the cooling and reuse of the hot water sent from the condenser, eliminating the possibility of thermal pollution caused by dumping the hot wastewater directly into a body of water such as the Susquehanna River.

Two workers inspect the 40-foot-tall louvers inside of one of the cooling towers for the Unit 2 reactor in this image taken on October 22, 1974. During normal operation, heated water is pumped from the condenser into the top of the of the baffles—the louvered section of the cooling tower. The water cascades down the louvers and is naturally cooled by the updraft created within the structure. At the base of the cooling tower is a wide pool of water that evaporates into the updraft, further cooling the water cascading through the louvers. The evaporating water rises up through the cooling tower and leaves the structure as a cloud-like plume. The water that does not escape through the top of the cooling tower is reused and pumped back to the condenser. Water from the Susquehanna River is pumped into the system to make up for the water lost as it rises out through the cooling towers as vapor.

The image at right, taken while Unit 2 was still under construction in October 1974, illustrates the size of the louvers on the base of the cooling towers in comparison to an adult male. Approximately 400,000 gallons of water per minute flow through the cooling towers and are recirculated back to the condenser. In the below image, taken on February 29, 1980, after the nuclear accident (while both Unit 1 and Unit 2 were not in operation), the pool of water at the base of one of the cooling towers is visible as members of the media take a tour of the dormant power plant.

This is an aerial view of a few key buildings that comprise the Unit 1 and Unit 2 nuclear power plants. The cylindrical buildings in the center of the image are the containment buildings, with Unit 1 above, or to the north of, Unit 2. The respective turbine buildings are located next to each containment building, with the Unit 1 turbine building located to the right. The Unit 2 turbine building is located at the south side of the Unit 2 containment building. The electrical substation is shown at right (between the cooling towers).

A water sphere about half as tall as one of the cooling towers is positioned next to the cooling towers of Three Mile Island Unit 1 on the northern portion of the island. The water sphere, which is sometimes described as having an appearance similar to that of a golf ball sitting on a tee, was used as an emergency water source in the event of a fire.

The Three Mile Island observation center opened for public use in 1970. The center was built on a two-acre parcel of ground located along State Route 441 and facing the nuclear power plant. The grounds around the observation center were laid with a picnic area to provide a parklike setting for tourists and school groups. The building featured an observation deck, conference rooms, and educational and multimedia exhibits that provided an array of educational programs on topics ranging from the history of the island to the basics of nuclear power. Above is a view of the landscaped grounds and picnic area located behind the observation center. At right, Metropolitan Edison vice president John Herbein points out features in the scale model of the Three Mile Island nuclear power plant that could found in the observation center.

25

During the flooding that resulted from Tropical Storm Agnes in June 1972, the southern end of Three Mile Island, which is largely wooded and undeveloped, was mostly underwater. Above, local residents gather near the southern access bridge to Three Mile Island, watching as it collects debris and perhaps wondering if and when the bridge might wash away. Both the southern and northern access bridges to the island were flooded, cutting off access to the island. At left, the cooling towers for Unit 2 are nearly surrounded by floodwaters as the Susquehanna River crested at 32 feet on June 24. The northern portion of the island, where the nuclear power plant was being constructed, was largely saved from major flooding by a partially completed floodwall. For more information, see *Tropical Storm Agnes in Greater Harrisburg*, also published by Arcadia Publishing.

The Nuclear Regulatory Commission defines the control room as "the area in a nuclear power plant from which most of the plant's power production and emergency safety equipment can be operated by remote control." The complexity of the control room can be seen in its vast array of instruments that number well over 2,000. The control board, itself teeming with knobs, dials, and gauges, can measure 40 feet in length. As a result, control room operators at Three Mile Island were required to pass an extensive oral and written essay–based exam that tested applicants on everything from reactor theory to operation, controls, and safety systems.

The complexity of the control room was seemingly tempered by engineers who designed the plant to be run virtually by itself. As a result, prior to the accident in March 1979, many of the control room operators were not given sufficient training about how to respond to these system failures, and consequently, mistakes were more likely to occur.

In a picture taken on April 12, 1978, an employee of Burns International Security Services stands guard in the anteroom to the personnel air lock for the Unit 2 containment building where the nuclear reactor is housed. At the time, federal regulations required guards to not only be armed but also to be able to have the numbers to meet a forced intrusion of the plant with appropriate force.

Above is a detailed view of the control rod drive platform for the Unit 2 reactor. Below, the control rod drive platform rests on top of the reactor vessel head. The reactor vessel head is attached to the top of the reactor core, which is not visible here. Both of the nuclear reactors at Three Mile Island are pressurized water reactors. When the control rods are retracted from the reactor core and away from the fuel rods, nuclear fission, created from the uranium fuel, heats the water in the reactor core to a temperature of 600 degrees. Under high pressure of 2,200 pounds per square inch, the water cannot convert to steam. This water is circulated through the closed primary loop from the reactor to the steam generators, where it transfers heat, not radioactivity, to the water in the secondary loop.

While the primary loop of the power plant is located entirely within the containment building due to radioactivity within the water, this is not the case with the secondary loop. The water within the secondary loop that is heated within the steam generators is converted to steam under high pressure and it is piped from the containment building to the turbine building. Above, plant technicians work near one of the instrument panels within the basement of the Unit 1 turbine building. There are two turbine buildings on Three Mile Island, one for each nuclear reactor. At left is a small portion of the ductwork found in the basement of Unit 1, which carries steam to the turbine building.

The pressurized steam in the secondary loop is sent from the steam generator in the containment building and travels through ductwork in the basement of the turbine building up to the turbine. The pressurized steam is forced across the blades of the turbines, spinning the blades of the turbines at a rate of 1,800 revolutions per minute. The turbines then turn the generator, which ultimately produces electricity that is sent to the transformers in the adjacent substation. Above, station superintendent Gary Miller stands within the upper level of the turbine building for Unit 2 and points toward the high-pressure turbine, which receives the steam from the steam generator. Below is an overhead view of the Unit 1 turbine generator while it is receiving maintenance (and not in operation) in April 1978. The high-pressure turbine, along with one of the three low-pressure turbines, is shown uncovered.

Once the steam has passed through the turbine, it moves through a condenser, where it is cooled and condensed back to liquid (water) so that it can be cycled back to the steam generators. A separate source of water to cool the condenser is cycled through the cooling towers. Above is a view of the inside of the Unit 2 turbine room as it was on February 29, 1980. Below is a view inside the basement of the turbine building of Unit 2 with a plant technician standing next to the feedwater heater. The feedwater heater raises the temperature of the water in the secondary loop prior to it being sent back to the steam generator.

Approximately every 12 to 18 months, approximately one-third of the 177 fuel assemblies in the reactor become spent and can no longer sustain a nuclear reaction. The reactor vessel head is removed to gain access to fuel assemblies. Above, Jack Herbein, vice president of generation for Metropolitan Edison, explains the fuel inventory of the Unit 2 reactor and the spent fuel pools to members of the press during a tour of the plant on April 12, 1978. At right, a crew working inside of the Unit 1 containment building utilizes the polar crane and fuel handling bridge during the refueling process. The process of refueling each reactor takes, on average, six weeks to complete.

The spent fuel is removed and placed into pools of water inside the fuel handling building called spent fuel pools, as shown above. The fuel, although not useful in the reactor, still emits radiation and heat. The water in the pools cools the hot fuel and provides a shield against any harmful radiation. Typically, the pools will be up to 40 feet deep to provide adequate shielding and also to accommodate the long fuel assemblies. The fuel assemblies are placed into racks to facilitate efficient storage. The racks also contain neutron-absorbing plates to minimize any nuclear fission. The fuel, on average, will spend five years in the spent fuel pool before it can be placed into casks for dry storage.

Two

THE ACCIDENT

Shortly before 4:00 a.m. on the morning of Wednesday, March 28, 1979, residents of the communities lining both the eastern and western shores of the Susquehanna River on either side of the Three Mile Island power plant were awakened by a loud roar, one far louder than those usually heard coming from the plant. The closest residents noted that the windows of their homes shook from the force of the sound. That force would soon shake the world.

Several mechanical failures and human errors contributed to the nuclear accident on March 28, 1979, at Three Mile Island. The first error occurred just before 4:00 a.m., when workers attempted to clear a blockage in the filters that cleaned the non-nuclear secondary loop. An errantly connected hose pushed pressurized water—instead of air—into the system, and the water worked into the air lines. This caused the valves for the feedwater system to slam shut and ultimately led to the

failure of the feedwater pump that sends hot water from the condenser to the steam generators in the secondary loop of the plant. The steam generator automatically shut down when it sensed that it was no longer receiving water. To alleviate pressure, a blast of steam was released from the turbine building; this was the noise that awakened area residents. As a result, water pressure and the temperature in the reactor's core began to rise.

A few seconds after the accident began, the Pilot-operated Relief Valve (PORV) automatically opened to release excess steam from the primary loop to the containment tank. Eight seconds after the accident began, control rods were lowered into the reactor, or scrammed, to slow the chain reaction and lower the temperature in the core. Pressure in the reactor began to fall, and the indicator light went out on the control panel. It was believed that the PORV had closed, but it had remained stuck in the open position. Emergency feedwater pumps were activated but unable to provide cooling water, as the valves in these pumps were closed a week earlier during maintenance procedures and not reopened. Steam and coolant were being lost from the primary loop, and the temperature was rapidly rising. Within the span of nine seconds, a Loss of Coolant Accident had begun.

At two minutes into the accident, the reactor's emergency cooling system pumps were activated to provide water to the reactor core during the Loss of Coolant Accident. It is then believed that the water levels within the core were safely rising and the pumps were then turned off, although, in reality, radioactive water and steam were still being lost through the open Pilot-operated Relief Valve (PORV). At eight minutes into the accident, control room operators finally opened the errantly closed valves for the secondary water loop to receive water from the feedwater coolant pumps that had been activated nearly eight minutes earlier. The emergency coolant pumps were also turned back on. Pressure in the reactor stopped falling, and there was sufficient coolant in the reactor to cover the fuel rods. All was stable for nearly an hour. Then the emergency coolant pumps began to vibrate and shake.

Amid fears that the problem was with the pumps and not the cooling system, the pumps were turned off. Without a steady supply of coolant reaching the core, and with the Pilot-operated Relief Valve (PORV) stuck in the open position, the water level in the reactor dropped, and the fuel rods were uncovered. The temperature in the core skyrocketed. The zirconium cladding covering the uranium fuel pellets ruptured, and the pellets began to melt. Radioactivity from the exposed fuel was released (as steam and hydrogen) into the coolant and out of the reactor though the open PORV into the containment building. Two and a half hours after the accident began, the PORV was finally closed. Radiation levels were several hundred times above normal in the primary loop, and operators were still not sure if the core was covered by coolant or not.

Just before 7:00 a.m., an emergency siren sounded on the island. For some plant workers, it was the first notification of an emergency situation. Some workers scrambled to their cars to leave the plant before it was locked down. A few minutes later, the Dauphin County Civil Defense office was notified of the emergency. High radiation levels in the containment building were setting off alarms in the Unit 2 control room. The Nuclear Regulatory Commission's regional office in King of Prussia, Pennsylvania, was notified of the situation at 7:45 a.m., and the notification was relayed to the national headquarters in Washington, DC, by 8:00 a.m. White House staff was notified by 9:15 a.m., and by 11:00 a.m., Three Mile Island had vacated all nonessential staff. Below, all vehicles leaving the island on March 28 were checked with a Geiger counter for elevated radiation levels.

"I can't talk now, we've got a problem," was the response that Mike Pintek, news director for WKBO radio received when he called the Three Mile Island control room around 8:00 a.m. on Wednesday, March 28. A traffic reporter for the station had overheard on a CB radio that state police had been mobilized to Three Mile Island and relayed the information to Pintek. WKBO broke the story on its 8:25 a.m. newscast. About the same time, the Associated Press in Philadelphia was relayed the information from a reporter in Waynesboro, Pennsylvania, who happened to be randomly checking in with the Pennsylvania State Police. At 9:06 a.m., the Associated Press released the story nationwide. In it, the Pennsylvania State Police conceded that a general emergency had been declared, "but there was no radiation leak." Soon, the press converged on the area.

Outside of the Three Mile Island observation center on Wednesday afternoon, March 28, Middletown mayor Robert Reid voiced his displeasure with the information supplied by Metropolitan Edison. Dauphin County Civil Defense notified Reid of the "on-site emergency" at 7:30 a.m., but it was not until 11:00 a.m. that a Metropolitan Edison spokesman called and assured him that there was no radiation leak. Moments later, a radio broadcast reported that there were indeed radioactive particles released.

"I wouldn't call it at this point a very serious accident. It's nothing we can't handle," commented John Herbein, vice president of generation for Metropolitan Edison, to reporters Wednesday afternoon on March 28 outside the Three Mile Island visitors center. However, at approximately 2:00 that afternoon, a thud was heard from inside the control room for Unit 2. The noise was a hydrogen explosion inside the containment building. It was largely ignored inside the control room.

On Thursday afternoon, March 29, Allen Ertel, a member of the House of Representatives from Pennsylvania whose jurisdiction covered Three Mile Island, attended a closed briefing with Metropolitan Edison officials and NRC inspectors inside the Three Mile Island Observation Center located directly opposite the island. Standing before a ring of reporters and camera crews, Ertel (above) summarized his feelings about the meeting by stating, "I don't think anyone can say they told us the absolute truth." Also in attendance at the meeting were Sen. Gary Hart (left) of Colorado, who served as chairman of the Senate Subcommittee on Nuclear Regulation, and Sen. Alan Simpson (right) of Wyoming, the ranking Republican of the subcommittee. Senator Hart commented as he left the meeting, "I think that it is much too early to make any judgment about this incident in connection with nuclear energy."

Although the previous day sent shock waves through the residents of Central Pennsylvania, life's daily routines went on as usual on Thursday, March 29. Under the direction of Vernon Lyter, a science teacher at Lower Dauphin High School, students were sent out into their surroundings to determine if any radiation could be detected as result of the accident at Three Mile Island. In the image at right, Kerry Hartman (left) and Steve Campbell, equipped with a Geiger counter, take readings from a rainwater puddle on the school's campus. Below, Jeanne Yingst (left) and Andrea Yannone are holding a scintillation detection device in order to measure the amount of radiation in the surrounding air. The students' efforts yielded results that indicated the radiation levels in the Hummelstown area, which is approximately eight miles north of the power plant.

"The business of the power companies is not to make power, but to make money . . . every dose of radiation is an overdose," exclaimed George Wald, Nobel Prize winner and professor emeritus at Harvard University (below right) at a press conference organized by Mobilization for Survival at the Friends Meetinghouse at Sixth and Herr Streets in Harrisburg. The event was organized by Rev. Robert Moore (below left) for the purpose of permanently shutting down Three Mile Island, and all nuclear power plants, due to their unsound safety practices. Ernest Sternglass, a professor emeritus for the radiology department at the University of Pittsburgh (below center), who gained notoriety in the 1960s for his research into the effects of nuclear fallout on children, commented, "It is not a disaster where people are going to fall down like flies . . . it is a creeping thing."

Above, John Herbein (left), vice president, and Walter Creitz, president, of Metropolitan Edison, field questions at a press conference at the Hershey Motor Lodge on Thursday, March 29. In attendance was Middletown mayor Robert Reid (right) who was still venting his frustration over the late delivery of information provided to his office the prior day. Conditions at Three Mile Island appeared to be stabilizing on Thursday, and Herbein commented, "There is presently no danger to the public health or safety. We didn't injure anybody, we didn't overexpose anybody and we certainly didn't kill anybody." Herbein did not mention to the press that on that same day, a "brief burst of radiation that measured 3,000 millirems" was measured by a helicopter flying near an outside vent. Furthermore, 40,000 gallons of slightly radioactive water was dumped from the plant into the Susquehanna River early Thursday evening.

On Friday morning, March 30, a Three Mile Island employee ordered a controlled release of radioactive gases from a water tank in the coolant system in order to relieve pressure in the system. The system was not airtight, and as a result, the radioactive gas was released into the atmosphere. Helicopters flying above the island soon reported readings of 1,200 millirems per hour. A statement from Three Mile Island officials owned up to the controlled release: "This release of gases is associated with efforts to handle gases in the primary coolant system." The press was also made aware of the 1,200 millirem-per-hour radiation levels, but poor communications within Metropolitan Edison's hierarchy left spokesman John Herbein hung out to dry when later that morning at a press briefing he stated that the release measured only in the 300 millirem per hour range.

Following the radiation release early on the morning of March 30, leadership at the Nuclear Regulatory Commission (NRC) initially recommend issuing an evacuation order for the general area but later backed off that request. Ultimately, after deliberating with his own state officials and the NRC, Gov. Dick Thornburgh held a press conference during the noon hour on March 30 announcing the decision to evacuate pregnant women and preschool children from within a five-mile radius of the plant. Twenty-three schools within this radius would also be closed. Residents within a 10-mile radius were advised to stay indoors and keep all windows and doors closed. Thornburgh stated that "this and other contingency measures are based on my belief that an excess of caution is best."

Soon after Gov. Dick Thornburgh issued the order to close the schools within a five-mile radius of Three Mile Island, a plan was put into place to evacuate Fishing Creek and Newberry Elementary Schools, along with Red Land High School. The students from these three schools, within the West Shore School District, would be bused to schools in the Northern York School District in Dillsburg, which is outside the 10-mile radius. In the above image, a mother picks up her son from Fishing Creek Elementary School before the remainder of the students are loaded onto the waiting buses lined up near the school entrance. In the below image, students wait in an auditorium of the old Northern High School for the arrival of their parents.

Searching for information, frustrated parents were met with busy signals when they tried to contact the schools and administration offices for information about the whereabouts of their children. Before the existence of websites, e-mail, and text messaging, dissemination of this critical information was through news broadcasts presented by local television and radio stations. Above, a mother picks up her two children from the old Northern High School. Below, high school–aged students receive directions outside of Northern Middle School. Red Land High School students who attended Cumberland Perry Vocational Technical School were the first to arrive.

Following the governor's order to close the schools within a five-mile radius of the power plant, a television camera crew films buses arriving at Middletown High School as a handful of students watch the spectacle in which they are also playing a part. Many of the schools within the 10-mile radius of Three Mile Island acted on their own accord and dismissed students early on Friday, March 30.

In this photograph, a family arrives at the Red Cross shelter at Hershey Park Arena carrying only the bare essentials. The shelter was touted as the "first shelter in the country as a result of a nuclear accident" and received approximately 150 individuals, mostly pregnant women and young children, following the governor's evacuation order on March 30.

The Red Cross staff did their best to keep the children at the Hershey Park Arena shelter busy with games and activities, even as members of the ever-present media hovered nearby. Although this shelter was initially opened for the pregnant women and young children affected by the evacuation order, the decision was made later that day to accommodate entire families as cots and bedding were brought in from the nearby Army base at Fort Indiantown Gap and food was provided by Herco to accommodate several hundred people. Rev. William Mielke, manager of the Red Cross shelter, stated to the *Carlisle Evening Sentinel*, "We aren't interested in breaking up families." Those who remained at the shelter on Saturday morning, March 31, were greeted by more than 300 members of the media from across the United States and around the world.

After the issuance of evacuation and warnings on Friday afternoon, March 30, Middletown Borough dispatched its small fleet of fire trucks to patrol the streets of the borough using public address systems on the trucks to notify residents to stay indoors and abide by the curfew that was to take effect that evening at 9:00 p.m. Above, a borough fireman speaks with a resident along the 100 block of Adelia Street who had come out onto the front porch of his residence. In the below image, Mayor Robert Reid confers with one of his borough employees who was equipped with a Geiger counter to measure and log any radiation that might be present in Middletown Borough on Friday, March 30.

In March 1979, Middletown Borough was home to approximately 11,000 residents. It was estimated, at that time, that 20 percent of the population packed up and temporarily left the area due to the uncertainty of what might transpire at Three Mile Island. Above, Mary Lee Shaffer (left) and Sally Ann Shaffer pack up the car outside of their residence on Oak Hill Drive in Middletown. Meanwhile, on the western shore of the Susquehanna River, it was estimated that up to 40 percent of the 500 residents of Goldsboro had packed up and left the area.

Oran Henderson (above) was serving as director of the Pennsylvania Emergency Management Agency (PEMA) during the accident at Three Mile Island. During the confusion on Friday, March 30, Henderson initially advocated for evacuation within a 10-mile radius of Three Mile Island, which was prompted by the radiation plume that was released earlier that morning. By the time any evacuation could be ordered, the radioactive plume had already traveled beyond the five-mile radius as it headed northeast. Henderson felt that at least the 20,000 residents within the five-mile radius could be evacuated within three hours. However, this estimate does not take into consideration the mass evacuation that would result from residents outside the five-mile radius who might be fleeing in panic. Below is a photograph of PEMA workers manning the radio room during the crisis.

The Department of Energy brought in a team of scientists to conduct aerial monitoring missions in the airspace surrounding Three Mile Island to measure and track airborne radiation plumes. Over 150 missions were flown during the days following the nuclear accident. Above, Dr. George Shipman (far left), holding Geiger counters, confers with colleagues and pilot Dick Eicher (far right) outside of one of the agency's helicopters. While the data collected during these flights was invaluable, difficulties remained in processing and transmitting the information in a timely fashion. By the time the information could be passed along to officials at the Nuclear Regulatory Commission, the airborne radiation could have traveled and dispersed across many miles. Below, a Department of Energy technician works within one of the mobile labs to process data received from monitoring flights.

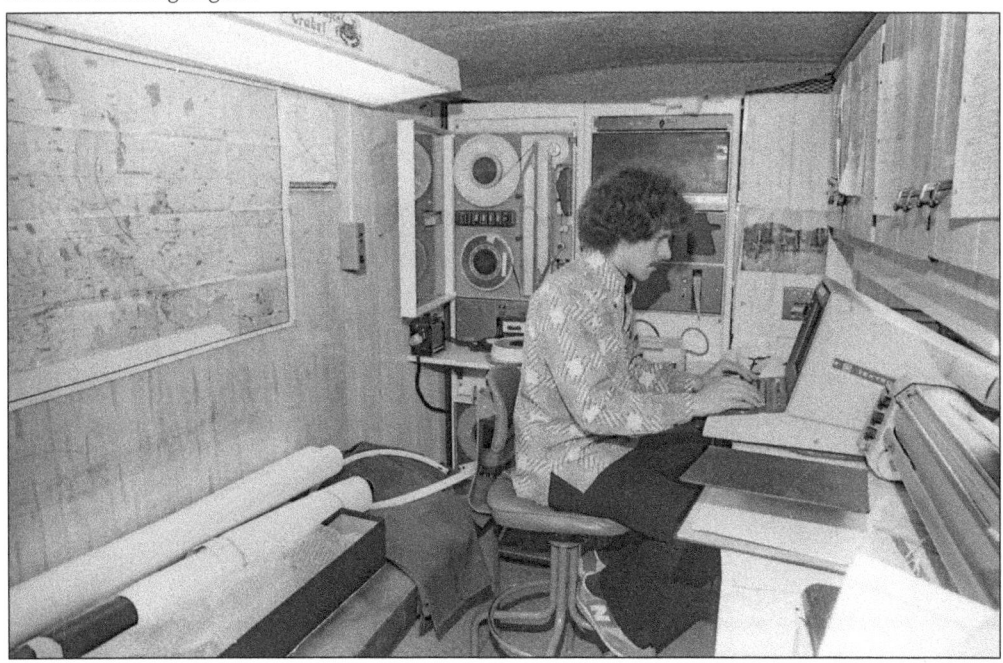

Gov. Dick Thornburgh not only found himself at the center of a crisis in which he lacked the scientific background to fully understand the situation at Three Mile Island without translation from experts in the field, but he also had to contend with widely differing opinions and interpretations that were being dispersed to his office and the media. Earlier in the morning on Friday, March 30, when Metropolitan Edison (Met Ed) vice president John Herbein was questioned about why state officials were not advised of the radiation release, he replied, "I don't know why we have to tell you each and every thing we do." The deteriorating situation with Met Ed prompted Thornburgh to request the NRC to send one individual who could speak with one voice on behalf of the government and Met Ed. They chose Harold Denton (below).

Harold Denton (above), trained as a nuclear engineer and employed by the Nuclear Regulatory Commission, was selected by Pres. Jimmy Carter to be the calming voice amidst the chaos. The decision to appoint Denton was timely, as the existence of a potentially explosive bubble had appeared in the reactor core. The bubble was partially composed of hydrogen and was believed to be 1,000 cubic feet in size. If the bubble increased in size, the fuel rods in the reactor would be exposed and no longer surrounded by liquid coolant. This could cause an increase in temperature and cause the hydrogen to either ignite or explode, allowing radiation to breach the core and containment building. The exposed fuel could also potentially cause temperatures to rise high enough to cause a meltdown, during which the hot radioactive materials literally melt through the core and containment building down into the earth and out into the environment.

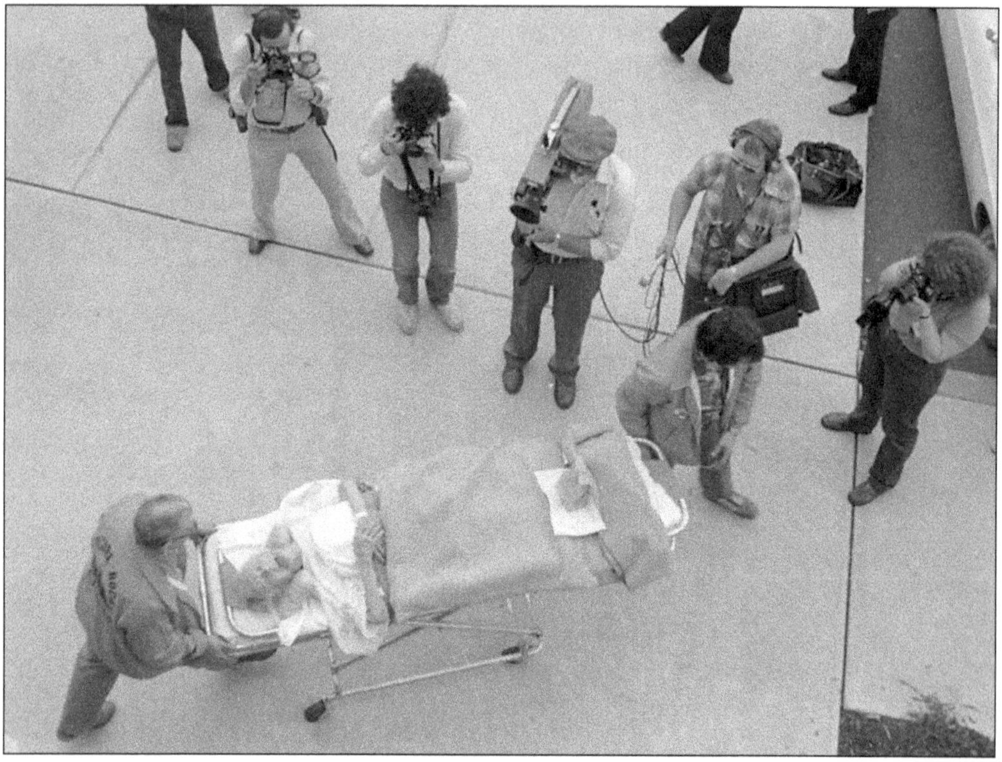

In these photographs, residents of Frey Village Retirement Home are being evacuated on Saturday, March 31. Two hundred and twenty residents were moved by ambulance, bus, and van from Frey Village to other facilities owned by Tressler Lutheran Service Associates. The decision to move the residents was made, in part, by officials of the rest home to try to stay ahead of the curve in the event of a mass evacuation. However, there were also concerns because the rest home was dealing with many staff members not showing up to work because they fled the area.

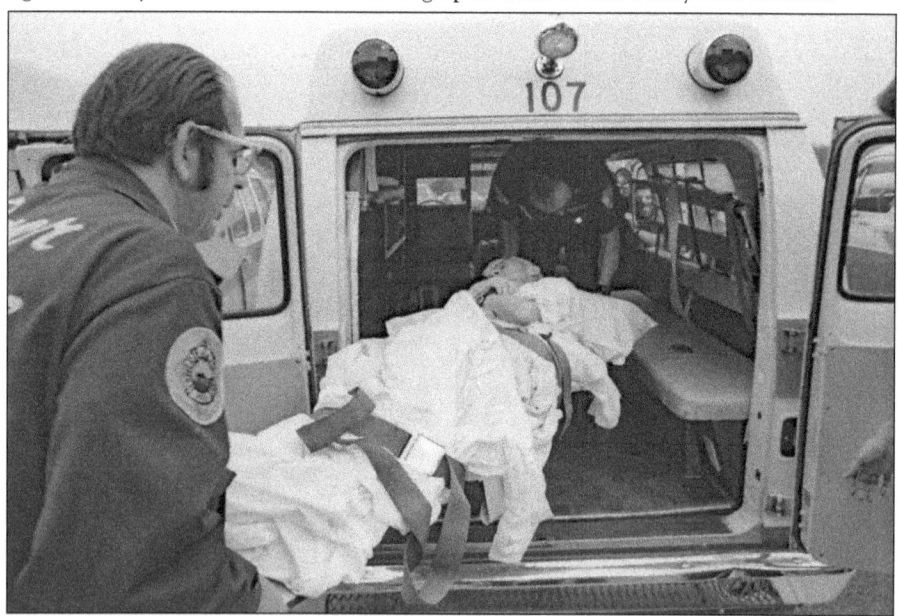

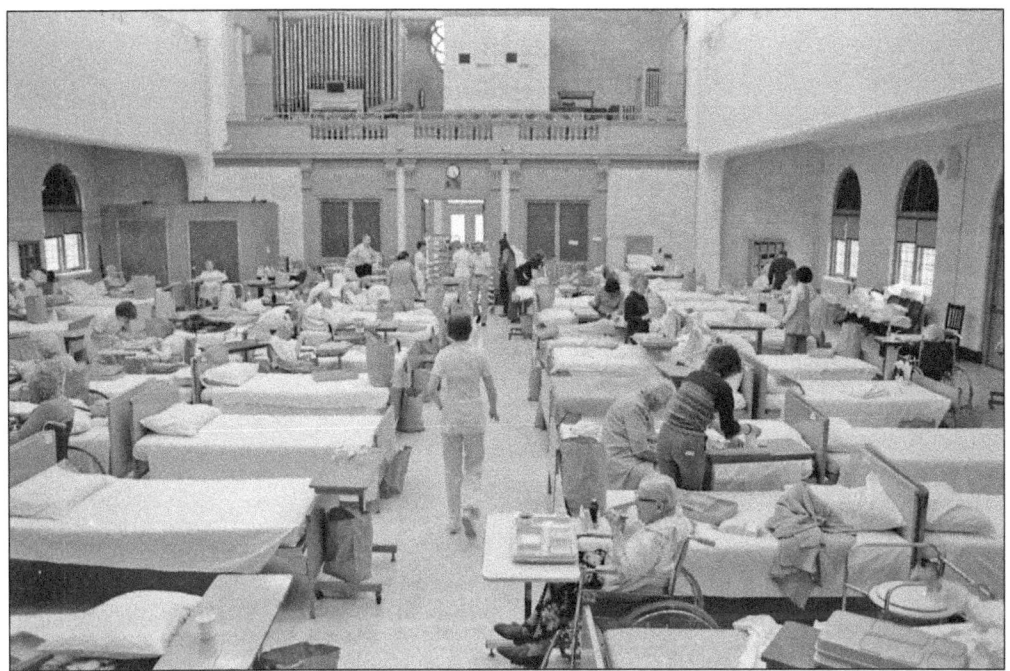

Within hours of the decision to evacuate the residents of Frey Village, officials at the Odd Fellows Home of Pennsylvania in Middletown followed suit and began preparations to move more than 140 of its residents. Arriving by ambulance and school buses provided by Central Dauphin School District, the residents were moved onto the grounds of the Pennsylvania State Hospital in Harrisburg on March 31. The majority of the residents were housed in the library and gymnasium. Those who were bedridden or required special care were housed in the chapel, as shown in these images.

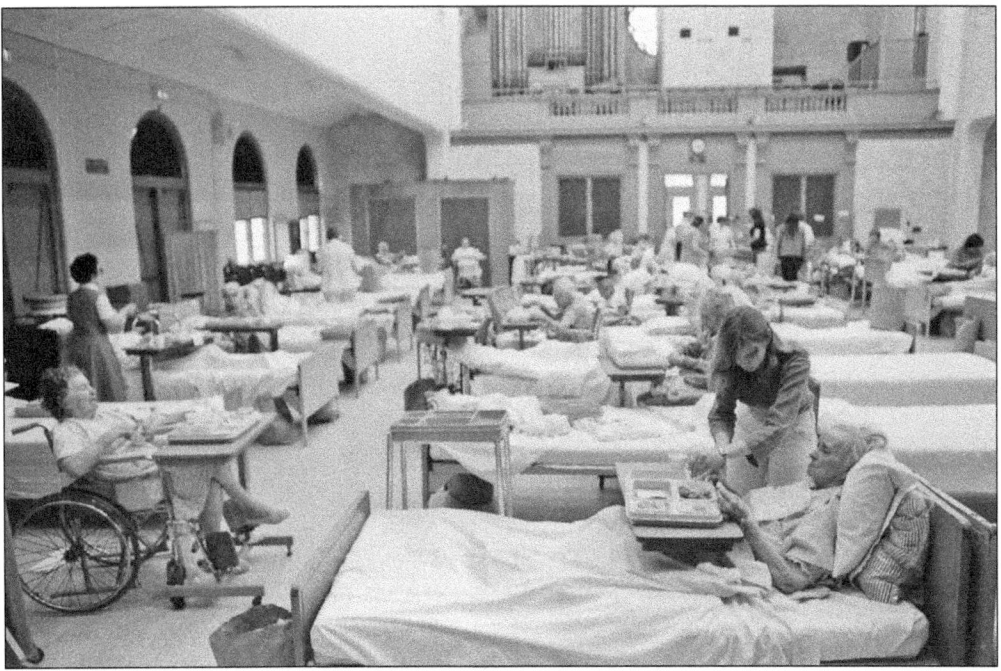

Word of the pending arrival of Jimmy Carter at Three Mile Island brought crowds, with most in good spirits. However, officials were still concentrating on bringing down the temperatures within the core of the Unit 2 reactor and achieving a "cold shutdown." Hindering this was the lingering bubble and new concern that had arisen late on Saturday, March 31. While the bubble's dimensions had decreased by 10 percent, the oxygen concentration within the bubble had increased when cooling water was pumped into the reactor at a faster rate. While the higher oxygen level raised the possibility of the combustion of the bubble, no explosion could occur without some form of ignition. NRC representative Harold Denton's statement to the press late on Saturday was to remain calm—"there is not (at this time) a combustible mixture and no near-term danger."

In the above photograph, Pres. Jimmy Carter and First Lady Rosalynn Carter arrive by Marine helicopter at the Air National Guard facility in Middletown, Pennsylvania, on Sunday, April 1, 1979. The president, who had served in the Navy as a nuclear engineer, was well versed in the technical language used in the operation and troubleshooting of a nuclear power plant. This prior experience served him well as he obtained daily briefings from his man in the field, Harold Denton. At the urging of Stuart Eizenstat, his domestic policy chief, Carter was to travel to Three Mile Island to provide reassurance to the locals—and the rest of the country—that any fear of danger had passed. In the below photograph, the presidential limousine approaches the security gate of Three Mile Island as the Secret Service keep pace on foot alongside the car.

Pres. Jimmy Carter and First Lady Rosalynn Carter, accompanied by Harold Denton and Gov. Dick Thornburgh, were transferred to a yellow school bus before being driven across the North Bridge to the nuclear power plant on Three Mile Island with a small coterie of the press trailing behind them. Once on the island, the president toured the control room of the Unit 2 reactor and was given a briefing about the current situation of the power plant. The tour of the plant was indeed brief, as the daily diaries of the president indicate that he spent no more than 13 minutes touring the plant before returning to the bus and ultimately back to the presidential motorcade, which then traveled a handful of miles to the city hall building of Middletown.

After his arrival at borough hall in Middletown, Pres. Jimmy Carter briefly met with his advisors before stepping inside a small gymnasium that was located within the facility. Standing before a small podium and flanked by Harold Denton to his right and Gov. Dick Thornburgh and Lt. Gov. William Scranton to his left, the president gave a brief statement to an audience that consisted largely of members of the press. His presence and his message were intended to calm the residents of the area and the outside world as he gave his reassurance that the situation at Three Mile Island was under control and steadily improving.

A crowd of spectators waits behind the rope barricade along West Emmaus Street, opposite borough hall, anticipating the arrival of Pres. Jimmy Carter. This was an extraordinary scene, given that the streets were largely deserted a few days earlier as thousands had fled from the area, and those who stayed behind were advised to remain indoors. Any anxiety or thought of danger seems to have been temporarily forgotten by the crowd waiting outside borough hall on that Sunday afternoon in April. As the president emerged from borough hall, he waved to the cheering crowds and stopped to give an autograph to a young boy before returning to the presidential motorcade. Although he spent no more than two hours on the ground in the Middletown area that afternoon, President Carter's presence seemed to signal to the local residents that things were going to be all right.

Inside the firehouse in Goldsboro, Three Mile Island and the ever-changing situation at the power plant garnered the attention of those who remained behind in the tiny riverside borough. Amidst all the gloom of the last five days, some good news was to be found on April 2, a Monday morning. Indications were received that the bubble had reduced by nearly 90 percent in the last day, although Harold Denton stressed some caution due to the rapid demise of the bubble with no great explanation for it. But when one concern seemed to be put into the rearview mirror, another one appeared. Milk samples collected within an 18-mile radius of the plant were found to contain trace amounts of iodine-131, which is a radioactive isotope with an eight-day half-life that can accumulate in the thyroid glands of humans.

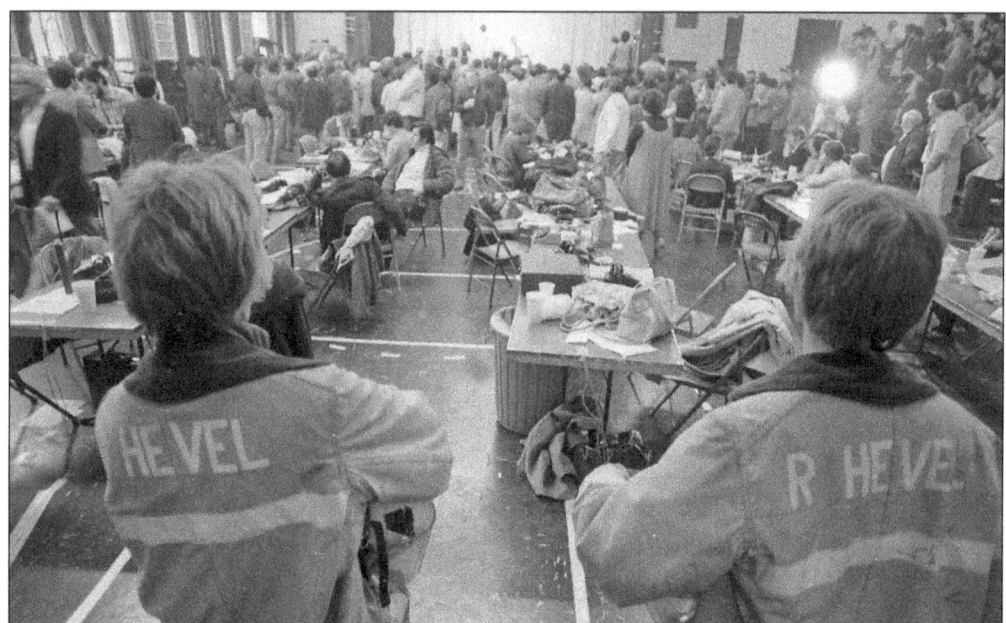

Harold Denton addressed the public and press corps at Middletown Borough Hall to assure them that the iodine levels were not a cause for concern. However, the public's cause for concern was that the airborne isotope would settle onto the grass in the pastures surrounding the power plant, and the dairy cows would ingest the iodine-131 isotope and pass it into the public milk supply. Some of the largest consumers of that milk would be children, who were also the group at highest risk. As a precautionary measure, the federal government ordered and received over 250,000 doses of potassium iodide that were housed in a warehouse near Middletown but never distributed. The iodine-131 leaks were eventually brought under control weeks later, when banks of charcoal filters were installed inside a venting stack that was leaking the isotope into the atmosphere.

On April 3, 1979, two NRC employees took readings with a Geiger counter to check for any signs of radioactivity along Route 441 in Londonderry Township with Three Mile Island and the Susquehanna River in the background. On this same date, Harold Denton advised the press corps that Metropolitan Edison had been dumping wastewater containing radioactive iodine and cobalt into the river until the NRC and DER requested they put an end to it.

Ten-year-old Melissa Stewart carries suitcases into her family's home on Nisley Street in Middletown. The family had returned home after a brief stay at a KOA campground in Jonestown, which is 30 miles northeast of Middletown. Some estimates place the number of individuals that evacuated from their homes at nearly 80,000 of the 200,000 residents from Dauphin and the other surrounding counties.

The Three Mile Island Observation Center, located along Route 441 in Londonderry Township and directly opposite Three Mile Island, was used as a hub to disseminate information to the press corps in the days following the nuclear accident. In the days and weeks that followed, the observation center would become the hub around which a growing encampment of more than two dozen tents and trailers, dubbed Trailer City, soon appeared. Space was cleared on the adjacent grounds for a helipad (above) to accommodate the helicopters that flew missions monitoring the release of radioactive airborne particles. As press briefings became a daily necessity, the major television networks set up their mobile transmission stations within the encampment (below) to provide audiences with up-to-the-minute coverage.

Trailer City was a self-contained working community that existed for the sole purpose of providing off-site support to the efforts of handling the aftermath of the Three Mile Island nuclear accident. The encampment allowed both the NRC and Metropolitan Edison to work together and coordinate their efforts while handling obstacles presented by the constantly evolving emergency. The small, self-contained working community had separate trailers for mobile laboratories (below) to process field samples on-site, accommodate staff meetings (above) for units such as the NRC Environmental Safety Unit, a kitchen and dining area to handle the extensive work schedules and basic needs of the workers, security personnel, and administrative and clerical support staff. A newsletter named the *Trailer City News* was written and distributed among the workers and provided everything from press clippings to parking arrangements to sports news within its mimeographed pages.

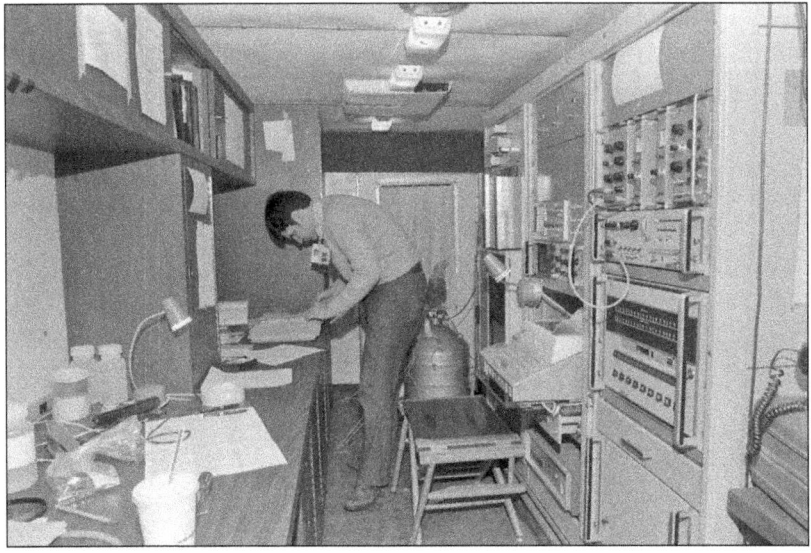

Middletown residents Jeffrey Paules (left) and Ronald Pristello (right) attend the daily press briefing in the gymnasium at the Middletown Borough Hall on April 4, 1979, where the NRC blamed the nuclear accident on human and mechanical failures, including two safety valves that were mistakenly left closed weeks before the accident.

Gov. Dick Thornburgh, along with his wife, Ginny, took telephone calls at the Governor's Action Center located at 900 Market Street in Harrisburg. Thornburgh felt that one of the largest problems facing the public after the nuclear accident was "the unending flow of rumors hurled at us from a variety of sources," and the call center would serve as a "24-hour citizens' center so the people can get the facts."

Above, students wait for a school bus on April 10, 1979, in anticipation of their first day back at school since March 30, 1979, when schools within a five-mile radius of Three Mile Island were dismissed under the recommendation of Gov. Dick Thornburgh. In the below image, students run toward the entrance of Grandview Elementary School for the first day back to classes on April 10, 1979. In comparison, the schools outside of the five-mile radius were back in session after being closed for only two days. However, many of the schools between the five- and ten-mile radius of Three Mile Island experienced low attendance rates. Fishing Creek Elementary School in York County, which was just beyond the five-mile radius, had an estimated attendance of only 25 percent of its pupils.

In these images, Harold Denton of the Nuclear Regulatory Commission addresses the press and public at a briefing held in the gymnasium within Middletown Borough Hall on April 4, 1979. Denton was brought in by Pres. Jimmy Carter on March 30 to handle the crisis at Three Mile Island when the threat of a meltdown and exploding bubbles seemed like real possibilities. Denton provided the press and the general public with a calm and assured voice that stated that while all was not well, there was a plan or plans in place to get them to safety. On April 17, 1979, with the threat of the bubble passed and the core's temperature cooling toward a safe level, Denton returned home to Bethesda, Maryland, a lengthy 19 days after he arrived at Three Mile Island.

The first-grade students of Miss Wolf's class at John C. Kunkel Elementary School hold a copy of a letter that they sent to Harold Denton following the accident: "Before we left school we were very scared and frightened. We were afraid the bubble would pop. We wanted to go far away because our houses are too close and we would have been very sick and maybe our houses would have exploded. Thank you for saying we could come home and for saving hundreds of lives. We love you for letting the air out of the bubble. We are glad it is over." Denton later responded to the children with a personalized video message thanking them for the letter. Kunkel Elementary School is located five and a half miles northwest from Three Mile Island and had the highest recorded radiation reading for an off-site location at 13 millirems per hour.

Residents who lived within a five-mile radius of Three Mile Island and fled due to the nuclear accident are shown waiting to file claims with American Nuclear Insurance and Mutual Atomic Energy Liability Underwriters at the Penn National building at Nineteenth and Derry Streets in Harrisburg. Claims were also accepted for pregnant women and families with preschool children within the five-mile radius based on the evacuation recommendation by Gov. Dick Thornburgh.

The sign outside of the Citizens Fire Company in Highspire, Pennsylvania, reads, "Keep the faith, we will survive." On April 28, 1979, the core of the Unit 2 reactor at Three Mile Island achieved a state of cold shutdown, during which the core will further cool by natural convection and without the help of coolant being forced into the core. At that point, the reactor is essentially in a state of dormancy.

Three

Fallout

"We Survived Three Mile Island," reads a homemade banner hanging from a Middletown porch on April 4, 1979. This phrase became a badge of honor for those who stayed behind and rode out the threats of meltdowns, radioactive plumes, and exploding hydrogen bubbles. As the Unit 2 reactor cooled, public activism began to heat up and take to the streets in opposition of nuclear power, and Three Mile Island was the rallying cry.

The first anti-nuclear rally in Central Pennsylvania following the nuclear accident at Three Mile Island was held on the steps of the Pennsylvania Capitol on April 8, 1979. The event was organized by Three Mile Island Alert and drew its share of seasoned anti-nuclear protesters. However, the rally also attracted locals who previously had no opinion about the power plant or nuclear power. Among those residents was Bruce I. Smith Jr. (below), a high school English teacher and chairman of the board of supervisors for Newberry Township in York County, who told the crowd, "They can make it into a national historical monument or Met Ed's memorial to be honest." After the accident, his wife, Patricia Smith, was appointed to the Newberry Township steering committee on Three Mile Island, which worked to ensure the power plant would not reopen.

In an effort to assuage the fears of locals, the Nuclear Regulatory Commission contracted with Helgeson Nuclear Services, Inc., to provide testing that included a full body scan of individuals, on a voluntary basis, to persons who resided within a five-mile radius of Three Mile Island. Several hundred residents volunteered for the scan at the mobile laboratory set up outside of Middletown Borough Hall. Testing consisted of lying down inside the scanner, which would check the subject's body for any radiation levels. The lab would then transmit the results of the test via telephone to the company's computers in California for evaluation.

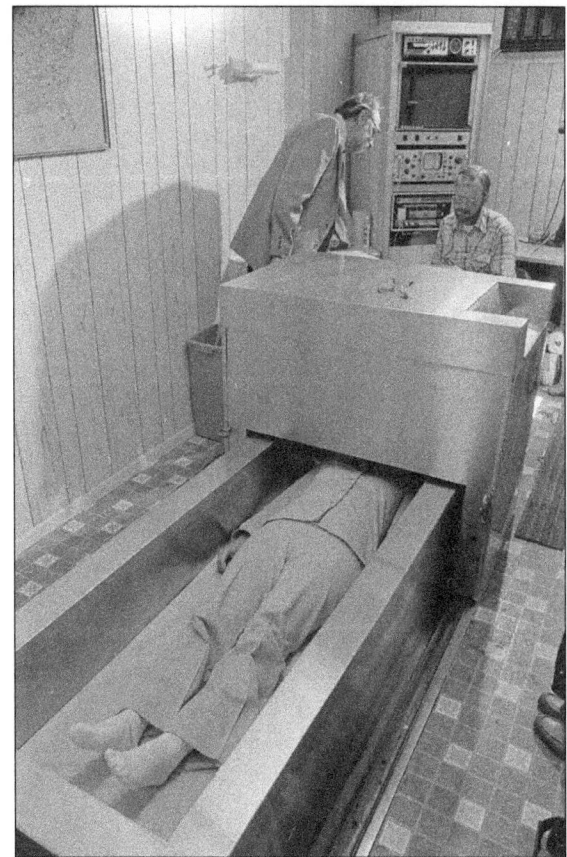

The tourism trade in Central Pennsylvania took a step backward in the weeks following the nuclear accident at Three Mile Island. This event negatively affected the attractions in Hershey and Pennsylvania Dutch country, which are located within a short driving distance from the nuclear power plant. While the negativity of the nuclear accident may have kept some tourists away due to fear of radiation exposure, some tourists found Three Mile Island a curiosity to be explored. Disaster (or dark) tourism is defined as traveling to places historically associated with death and tragedy. The attraction to Three Mile Island as a tourist destination was rather unique in the immediacy with which tourists arrived after the accident and could be tied directly to the amount of media exposure dedicated to the nuclear accident.

"I think Three Mile Island is becoming a tourist attraction of its own. Pennsylvania's got more visibility than it has had in a long time. Visibility is the name of the game when it comes to the travel industry," stated Pennsylvania commerce secretary James Bodine in the weeks following the nuclear disaster. Parked cars crowded the narrow spaces along the sides of Route 441 between Three Mile Island and the observation center in order to stop and take a quick picture with the cooling towers in the background. Souvenir vendors at roadside stands sold T-shirts proclaiming things like, "Hell No, I Don't Glow," and "I Survived Three Mile Island."

The President's Commission on the Accident at Three Mile Island was created on April 11, 1979, to insure "the safety of our citizens is never again endangered in this way," and tasked with "discovery of the truth" regarding the nuclear accident at Three Mile Island. The 12-member commission was chaired by John Kemeny (above), president of Dartmouth College, who, as a mathematician, worked on the Manhattan Project at Los Alamos National Laboratory. Other notable members of the commission included Bruce Babbitt, governor of Arizona; Harry C. McPherson, an attorney who served as special counsel to Pres. Lyndon Johnson; and Anne D. Trunk (below), a resident of Middletown. Trunk felt that she belonged on the commission as she "stayed in Middletown. They thought I could represent Middletown and people's feelings." The group was given six months to investigate, interview, and ultimately submit its findings.

The President's Commission held public hearings at Penn State Capital campus (above) and included testimony by, among others, experts within the nuclear field and locals residing near Three Mile Island. In an effort to obtain the truth, commission chair John Kemeny petitioned Congress for subpoena powers and the right to take sworn testimony from government and utility officials. The commission's finding concluded, in part, that "to prevent nuclear accidents as serious as Three Mile Island, fundamental changes will be necessary in the organization, procedures, and practices, and above all, in the attitudes of the Nuclear Regulatory Commission and, to the extent that the institutions we investigated are typical of the nuclear industry." While critical of the Nuclear Regulatory Commission and General Public Utilities, the purpose of the commission was never to recommend the demise of nuclear power but rather to insure a safer path for the industry.

Above, US House Reps. Morris Udall (left) and Allen Ertel field questions and pose for pictures following a tour held at the Three Mile Island power plant on May 7, 1979, for members of the Subcommittee on Energy and Environment. The tour was the first to allow elected officials into the crippled power plant since Pres. Jimmy Carter's visit on April 1. At left, 15 members of Congress tour the power plant outside the Unit 2 containment building, which houses the damaged Unit 2 reactor. The tour took the Congress members through the basement of the turbine room and ultimately into the control room.

While touring the control room, members of the congressional tour learned that the hydrogen explosion that had occurred 10 hours into the nuclear accident was known about by Nuclear Regulatory Commission inspectors inside the control room. However, this information was not relayed to their superiors in Washington until two days later. The hydrogen explosion was an indicator of damaged fuel inside the reactor, and had it been known earlier, it could have impacted decisions to evacuate.

John T. Collins (left), Harold Denton (center), and Harold E. Collins appeared on September 12, 1979, before the Pennsylvania House Select Committee on Three Mile Island. Harold E. Collins, director of the Nuclear Regulatory Commission's Office of Emergency Preparedness, testified that he ordered a mass evacuation within the five-mile radius on March 30, but Gov. Dick Thornburgh declined because he did not know who Collins was.

Metropolitan Edison employee David Klucsik (left) stands with an EPICOR II canister outside the Unit 2 auxiliary building. The EPICOR II system was designed to filter out radioactive strontium and cesium particles from Unit 2 coolant water. Below, Klucsik stands on the EPICOR II filtration canisters containing organic resin beads designed to trap the radioactive particles. However, the system was not capable of capturing tritium, a radioactive form of hydrogen. The initial plan was to release the filtered water with tritium into the Susquehanna River. A lawsuit from the City of Lancaster and the Susquehanna Valley Alliance was filed opposing the plan, as more than 100,000 residents downstream of the plant used the river as a source for drinking water. The lawsuit was ultimately dismissed, and the EPICOR II system began operation on October 18, 1979.

Metropolitan Edison spokesman Sandy Polon explains to the press the problems with testing the EPICOR II system: "the NRC won't allow us to test it with radioactive water." The $5 million system was designed to filter 400,000 gallons of radioactive water in three months. Due to additional passes through the filtration system and frequent resin changes, the process of filtering 500,000 gallons of radioactive water was not completed until August 9, 1980.

"Caution, high rad levels inside vault, radioactive materials, notify health physics before entering," reads the card attached to a concrete vault outside the Unit 2 auxiliary building. As a result of the nuclear accident, 600,000 gallons of radioactive water flooded the basement of the Unit 2 containment building. This water was decontaminated using resins of the mineral zeolite in a Submerged Demineralizer System. The process took six months and was completed in March 1982.

Jane Fonda and her husband, political activist Tom Hayden, speak to a crowd of 1,000 in the ballroom of the Penn Harris Motor Inn on September 23, 1979. Fonda and Hayden came to the Harrisburg area to speak out against nuclear power and kick off a 50-city nationwide tour to support their Campaign for Economic Democracy program. "As an actress . . . during the filming of *China Syndrome*, I had to imagine what it was like to be in a nuclear plant that was out of control. But, there was no way that I could imagine what it was really like. I'm so sorry that it happened to you and I hope it doesn't get any worse. My heart goes out to you."

Holly Garnish stands near her home on Meadow Lane, which is the closest in proximity to the Three Mile Island Nuclear power plant. Following the nuclear accident on March 28, 1979, the Nuclear Regulatory Commission placed radiation detectors in the shrubs surrounding the Garnish home. On March 29, she began to document her family's emotional state in a daily diary: "kids are bordering on panic . . . I can't seem to pray, leave it in God's hands."

George Kunder, shown here, served as a supervisor of the Unit 2 reactor in March 1979, and he was the first person to be called by the control room operators soon after the accident occurred. As he drove the short distance from his home to the power plant, he thought the situation would probably be under control by the time he arrived.

Biologist Roger Thomas kneels on the pier at the Fish Commission access area in Goldsboro on September 12, 1979, as he puts mussels into a wire basket that will placed in the Susquehanna River as part of a study to measure the amount of radioactivity absorbed by the mussels.

On December 6, 1979, members of the press viewed video footage that was obtained by inserting a video camera through a hole bored through the concrete wall of the containment building. These were the first views from within the building since the nuclear accident. Robert Arnold, vice president of Metropolitan Edison, announced that no visible damage to the reactor from the hydrogen explosion on the afternoon of March 28 was detected.

Local residents attended a meeting held in Middletown Borough Hall on December 13, 1979. The meeting was intended to present the topic of deep safe geological containment of nuclear waste. Speaking at the event were, from left to right, Harry C. McPherson, member of the President's Commission on the Accident at Three Mile Island; Dr. Frank Press, geophysicist and advisor to Pres. Jimmy Carter; Rep. Allen E. Ertel; and Stanley M. Gorrison, chief counsel for the President's Commission on the Accident at Three Mile Island. But, the subject of the meeting turned toward the nuclear accident and the role played by Metropolitan Edison, to which McPherson replied, "I thought that Met Ed were a bunch of turkeys . . . from my point of view, that they were to be faulted in their operation" and "their response once the accident occurred. I found it to be a totally unacceptable operation."

California governor Edmund Gerald "Jerry" Brown Jr. speaks to reporters at the Three Mile Island media center after a brief tour of the power plant. Brown had made the trip as part of a tour seeking support for the Democratic Party's nomination in the 1980 presidential election. "President Carter has become the chief salesman for the nuclear industry. But I do not think any of us should be buying his product."

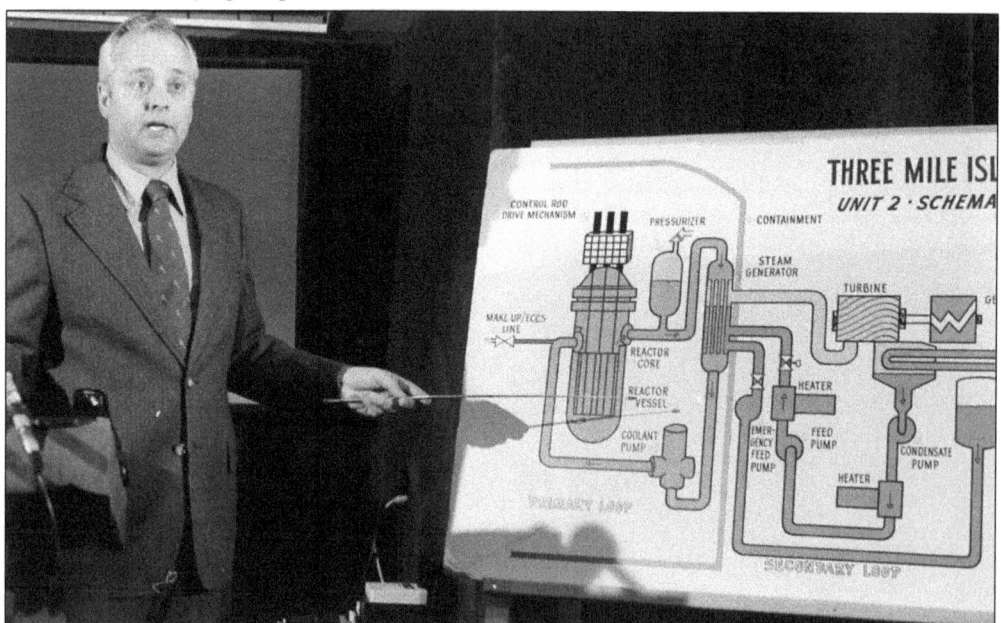

In this image, General Public Utilities vice president Robert Arnold addresses the press on December 28, 1979, regarding the presence of radioactive krypton-85 gas, which was detected outside the power plant. Metropolitan Edison sought permission from the NRC to vent the krypton gas in a controlled manner to avoid any future uncontrollable leaks that could possibly occur from the eventuality of failing equipment within the Unit 2 reactor building.

Activist Gene Stilp is shown giving an interview to the press during public hearings held at Elizabethtown Middle School on March 20, 1980. With several hundred people in attendance, elementary school teacher Thomas Bainbridge jumped onto the stage with General Public Utilities president Robert Arnold and asked if a meltdown was possible at Three Mile Island. Ultimately, before Bainbridge left the stage, Arnold admitted that a meltdown was a possibility.

Demonstrators gathered at 2 Meadow Lane in Londonderry Township a few minutes before 4:00 a.m. on March 28, 1980, for a candlelight vigil to mark the first anniversary of the nuclear accident at Three Mile Island. William Vastine, of the nuclear power opposition group Three Mile Island Alert, passed out candles, while those in attendance sang, "No more nukes, my lord, kumbaya, shut them down, my lord, kumbaya."

A crowd gathered at 2 Meadow Lane in Londonderry Township on Friday, March 28, 1980, for a "speak out" on nuclear power and Three Mile Island. Below, ecologist and author Barry Commoner addresses the gathered crowd of protesters and press at the "speak out" on Friday morning. At the rally, Commoner announced his bid for the 1980 presidential race as a candidate for the Citizens Party, which was founded in 1979 by Commoner for the purpose of addressing progressive, pro-science, and environmentalist issues. Commoner had gained notoriety in the early 1970s for his book *The Closing Circle*, which introduced sustainability of natural resources to the general public.

The National Nuclear Debate was broadcast live on the Public Broadcasting Service (PBS) from the Forum building in Harrisburg on March 28, 1980. Below, program moderator Jim Lehrer speaks to the crowd prior to the beginning of the debate in an effort to control a crowd that was largely anti-nuclear and quite vocal cheering and booing the responses of the participants. Dr. Norman Rasmussen (seated at far right), a native of Harrisburg and head of the nuclear engineering department at MIT, defended nuclear power and the problems at Three Mile Island as the growing pains of the industry. Dr. Henry Kendall (seated at far left), who was awarded the Nobel Prize in physics in 1990 and chaired the Union of Concerned Scientists, stated, "I don't think it is worthwhile making a laboratory out of the United States so these people can deal with the problems."

The largest of the many events held to commemorate the one-year anniversary of the accident at Three Mile Island was a rally on the east side of the Pennsylvania Capitol on March 29, 1980. The event was organized by the anti-nuclear and environmentalist collective March 28 Coalition, named after the date on which the Three Mile Island nuclear accident occurred the prior year. A crowd of between 5,000 and 7,000 attendees braved the cold and rainy March afternoon. Although the focus of the event was the remembrance of the events that occurred at Three Mile Island the previous year, many environmental activists used the opportunity to speak out against nuclear power in general.

The event was highlighted by performances from musicians who lent their voices to the antinuclear cause, including John Hall (above), a founding member of the band Orleans, known for their single "Still the One." In 2006, Hall was elected to the US House of Representatives as a Democrat in the traditionally Republican New York Nineteenth Congressional District. Below, pictured left to right are Kenny Edwards, Wendy Waldman, Linda Ronstadt, and Pete Seeger, who came together to sing the Woody Guthrie standard "This Land is Your Land" and the traditional American folk song "We Shall Not Be Moved." Seeger, who had once been a proponent of nuclear power as an alternative energy source, told the crowds, "the poor may inherit the earth, but they will inherit such a poisonous garbage dump that they probably won't want it."

Above, Robert Arnold, president of General Public Utilities, addresses the press on June 28, 1980, regarding a planned release of krypton-85 radioactive gas from the Unit 2 reactor in a measure approved by the Nuclear Regulatory Commission. Over 50,000 curies of the krypton gas remained within the Unit 2 containment building; it was planned to be gradually released in order to allow workers to continue with the cleanup process. To illustrate the safety of the venting process, a picnic was organized by the organization Friends and Family of Three Mile Island; family members of the plant's workers were encouraged to attend and partake in food and games (left). However, the initial venting only lasted for a few minutes as alarms sounded when excess radioactive particles were detected, which was explained as an initial rush of krypton gas before flow was stabilized.

At right, protesters march across the Market Street Bridge from City Island in Harrisburg on March 28, 1981, to mark the second anniversary of the nuclear accident at Three Mile Island. What made the protests different on this day was the makeup of those who chose to participate in the marches and rallies. The United Mine Workers of America was one of the organized groups that joined the protest, as the group viewed nuclear power as a threat, albeit for different reasons. Below, anti-nuclear protesters with banners in hand march up Second Street toward the Pennsylvania Capitol.

As several thousand protesters arrived at the eastern side of the Pennsylvania State Capitol Complex on the second anniversary of the nuclear accident, the message was no longer confined to issues directly related to Three Mile Island. As many as nine organized labor groups joined the protests, and while nuclear power was a common thread in their messages, each group had something different to say. The United Mine Workers of America protested nuclear power not as a threat to the fossil fuel industry but as a stepping stone to nuclear weapon proliferation. At the forefront was Barry Commoner, who aimed to draw common alliances between the anti-nuclear groups and the labor unions. "From this day forward, I can tell you the environmentalists and the anti-nuclear alliances will march side by side with the unions."

In this picture taken on February 10, 1982, Robert Arnold, president of General Public Utilities, points to a model of the steam generator—specifically, to the heat exchange tubes of the Unit 1 reactor. Approximately 1,200 of the 31,000 heat exchange tubes became corroded and cracked due to sulfur in the system. The heat exchange tubes carry superheated water into the steam generator to create steam, which is then sent to the turbines.

Unlike in previous years, the third anniversary of the nuclear accident at Three Mile Island was much smaller and more intimate. No large marches or rallies were held. Only a small group of individuals braved the cold to hold a candlelight vigil outside the power plant. A church service and a chicken corn soup dinner were other events held to commemorate the anniversary.

On May 18, 1983, protesters stand outside the gates of Three Mile Island on the one-year anniversary of the referendum vote taken in Dauphin, Cumberland, and Lebanon Counties, where, by a margin of two to one, residents voted against the restart of the Unit 1 reactor. However, the referendum was nonbinding and held no real authority to keep the Unit 1 reactor from restarting. What kept the Unit 1 reactor out of commission, at that time, were corroded heat exchange pipes and an order from the US Court of Appeals that a psychological impact assessment on local residents must be completed before consideration would be given to the restarting of the reactor.

On August 30, 1983, a cask carrying the last shipment of the "abnormal radioactive waste," filtered from the highly radioactive wastewater from the Unit 2 accident, left aboard a flatbed truck bound for Richland, Washington. Prior to the waste's departure, General Public Utilities president Robert Arnold told those in attendance, "We recognize our neighbors' concern, and we are committed to Three Mile Island not becoming a permanent waste disposal site."

In this image from December 12, 1983, Jack DeVine, technical planning director for Three Mile Island, shows video footage of the damage within the core of the Unit 2 reactor. The video imagery offered an early glimpse at the fractured fuel assemblies and the portion of the core that collapsed into rubble. Sonar mapping showed that a portion of a three-quarter-inch steel plate within the reactor vessel had buckled during the accident.

In these images, technicians are employing different methods of scabbling, the manual or mechanical removal of a thin layer of concrete or other hard surface that has been contaminated by radioactive particles. The scabbled material is then vacuumed off and handled as radioactive waste. Scabbling is performed with either a jackhammer (above) or by a specially designed machine (below) that is capable of abrading a concrete surface and vacuuming away the particulate matter. Although it is a labor-intensive means of decontamination, the end result produces far less physical waste that will need to be shipped to and ultimately stored at a nuclear waste disposal facility.

Above, Edwin Kinter, executive vice president of General Public Utilities, shows the members of the press an illustration of the Unit 2 reactor core prior to the partial meltdown. This was part of a tour held on March 21, 1984, to mark the fifth anniversary of the nuclear accident. Kinter, a former employee of the Department of Energy, had been placed in charge of completing the cleanup process of the Unit 2 reactor. At right, as part of the fifth anniversary tour, members of the press were given an opportunity to photograph and film some of the interior spaces, including the inside of the command center within the Unit 2 control room, as shown here.

A crowd of several hundred gathered in the rain opposite the Three Mile Island nuclear power plant just before 4:00 a.m. on March 28, 1984, for a candlelight vigil in remembrance of the nuclear accident that occurred five years earlier. The normally low-key event soon garnered more than the usual media attention when word spread that Rev. Jesse Jackson would be in attendance. Jackson was seeking the Democratic nomination in that year's presidential election, and his stop at Three Mile Island was shortly before the Pennsylvania Democratic primary was to be held on April 10, 1984. Jackson's motorcade (above) was greeted with cheers, and he briefly spoke to the crowd: "For five long years, you have lived in the shadow of this plant and the shadow of the accident. We are here today to oppose the restart of Three Mile Island."

As part of a cross-state tour two days ahead of the Pennsylvania Democratic primary, former vice president Walter F. Mondale scheduled a stop outside the gates of Three Mile Island on April 8, 1984. The issue of whether or not to restart the Unit 1 reactor became one of the touchstone issues for Democratic candidates in that year's presidential primary, and the 100 or so locals who gathered at the plant that day were highly receptive to Mondale's message: "The facts speak for themselves. Based on what we now know, Three Mile Island should not be allowed to open."

John Anderson, who ran as an independent candidate during the 1980 presidential election, is shown campaigning for Democratic presidential candidate Walter Mondale outside of the Three Mile Island on October 2, 1984. Anderson was critical of Pres. Ronald Reagan appointing individuals to the Nuclear Regulatory Commission who were apt to approve a restart of the Unit 1 reactor instead of looking out for the interests of the locals.

Speaking at an anti-nuclear rally on the eve of the fifth anniversary of the accident at Three Mile Island, political activist Ralph Nader is shown addressing a crowd of 1,000 at the Forum building in Harrisburg on March 27, 1984. As the main attraction of the night's events, Nader advocated for activism on an individual and local level. "Every major social movement in this country's history started from citizen action."

As an act of civil disobedience, protesters gathered on May 18, 1984, to recognize the second anniversary of the local referendum vote to prevent the restart of the Unit 1 reactor. In the above image, a banner alluding to the upcoming 1984 presidential election is carried along Route 441 toward the gates of Three Mile Island. Below, eight protesters form a chain across the entrance to the north gate of Three Mile Island. When they refused to peacefully disperse, they were taken into custody by members of the Pennsylvania State Police. In all, 50 individuals were charged with obstructing a public highway, with one person charged with disorderly conduct for fighting with police.

Above, Clarence "Swede" Hultman of the Bechtel Corporation is shown describing the head lift operation for the Unit 2 reactor in this image from July 18, 1984. Removing the 25-foot-tall, 156-ton steel head of the nuclear reactor is a routine operation performed during normal refueling operations. The reactor head was to be lifted by crane and ultimately placed on a stand where it would rest. A plastic covering, or "diaper," was placed on the bottom of the head to contain any residual radioactive particles. One technician was treated for radiation exposure when he removed his protective respirator during the head lift procedure. Below, Greg Eidem of General Public Utilities describes the head lift operation to the press through video recordings filmed during the operation that took place on July 24, 1984.

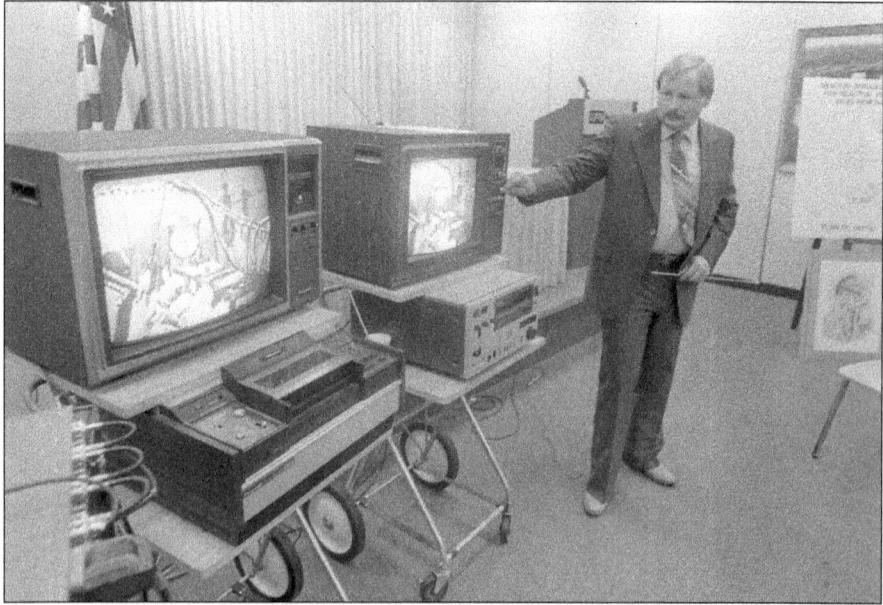

In the above image, a group of protesters with T-shirts spelling out "NO RESTART" marches along Route 441 toward Three Mile Island on February 24, 1985. Nearly 150 protesters took place in the march and rally outside the power plant to voice their opposition to the restart of the Unit 1 reactor. The protesters voiced their concern over the Nuclear Regulatory Commission's lack of hearings to be held ahead of their vote to decide whether Unit 1 could again become operational. Among the hearings not held were those that addressed alleged falsification of data related to Unit 2. At right, Ed Charles of Mechanicsburg stands in front of the Three Mile Island sign near the plant's north gate with a cowboy that looks conspicuously like Pres. Ronald Reagan.

On May 29, 1985, the Nuclear Regulatory Commission voted 4–1 to allow General Public Utilities, the owners of Three Mile Island, to restart the Unit 1 reactor. Federal court challenges were filed by Gov. Dick Thornburgh and Three Mile Island Alert. At issue, in part, was the Nuclear Regulatory Commission's "refusal to hold public hearings on management integrity and competency and safety issues creates the appearance of a cover-up," stated Sen. John Heinz. With details agreed upon between protest leaders and the Pennsylvania State Police ahead of the vote by the NRC, a nonviolent protest rally, which would simulate a wedding procession, was planned to march to the north gate of Three Mile Island, as shown above. Upon arrival at the gate, protesters blocked the gate in waves of as many as 20 at a time.

At 6:00 p.m. on May 29, 1985, a total of 250 protesters arrived at the north gate entrance to Three Mile Island. In groups of 20, they would stand at the gate and wait, in turn, as they were peacefully escorted—or carried away—by the state police; this was negotiated the previous day between protesters and the police. In all, 82 protesters were arrested and taken by bus to a special booking area set up in a hangar at nearby Harrisburg International Airport. Once there, they were charged with the summary offense of obstructing a public passageway and ordered to pay an amount of $73.50, which covered a fine and court costs. "The people said they wanted a nonviolent protest, and that's what they have done," stated Tom Lyon, spokesperson for the Pennsylvania State Police.

On June 7, 1985, Judge Collins Seitz of the Third US Circuit Court of Appeals ordered that Unit 1 must remain shut down pending a hearing on June 27. At this hearing, the Nuclear Regulatory Commission was ordered by the court to further investigate claims of mismanagement by General Public Utilities, the owners of Three Mile Island. However, on August 27, 1985, judges from the same Third US Circuit Court of Appeals ruled 2–1 to deny petitions by four groups, including the State of Pennsylvania, to block the restart of Unit 1 until further investigations of Three Mile Island's management could be further scrutinized. Hours after this ruling, Gene Stilp set up outside the north gate of Three Mile Island to protest the court's decision. He was arrested, as shown above and below, for blocking a public utility right-of-way.

In this image, protesters gather outside the Governor's Mansion in Harrisburg on August 28, 1985, to call on Gov. Dick Thornburgh to pull the state of emergency planning for Unit 1. Without Pennsylvania's cooperation on emergency planning, the power plant could not legally begin operations. Thornburgh refused to follow through with this tactic due to the damaged fuel that remained inside the Unit 2 reactor and might necessitate future evacuation.

On August 29, 1985, the State of Pennsylvania petitioned the appeals court to halt the restart of Unit 1, which was granted pending appeals of the prior June 27 ruling. However, on October 2, 1985, the US Supreme Court ruled in favor of General Public Utilities and the restart of Unit 1. In the above image, protesters stand in the rain along Route 441 opposite Three Mile Island following the ruling.

Here, Three Mile Island Alert member Eric Epstein is carried off by members of the Pennsylvania State Police following a protest of the US Supreme Court's ruling to allow the restart of the Three Mile Island Unit 1 reactor on October 2, 1985. The Nuclear Regulatory Commission approval of the restart of the Unit 1 reactor followed soon after the Supreme Court decision. Phillip R. Clark, president of General Public Utilities, commented, "The democratic process worked. There have been hearings. Everybody has had a chance to be heard." At 4:30 a.m. on October 3, 1985, the control rods in the Unit 1 reactor were lifted, allowing the uranium fuel to begin the process of nuclear fission.

Following the Supreme Court ruling and the NRC's approval of the restart of the Unit 1 reactor, Gene Stilp and a small group of protesters stood defiantly outside the Governor's Mansion in Harrisburg to voice their displeasure at Dick Thornburgh's concession that it was time to change tack and work with the NRC and General Public Utilities to insure the power plant was operated in a safe manner going forward.

At 1:30 p.m. on October 3, 1985, criticality was achieved within the Unit 1 reactor, which is a state of maintaining a self-sustaining nuclear chain reaction—a first in over six years. Phillip Clark, president of General Public Utilities Nuclear, advised the press that "everything is going smoothly. Our instructions to the operating crews have been to proceed deliberately and carefully, placing primary assurance on doing things safely."

At 4:02 a.m. on October 9, 1985, the turbine generator for the Unit 1 nuclear reactor began producing electricity as the Unit 1 reactor reached 15 percent power, the threshold for the nuclear reactor to begin producing power. It was the first time that the Unit 1 reactor had produced commercial electricity since it had been taken offline for refueling on February 17, 1979.

Frank Standerfer, director of the Unit 2 reactor for General Public Utilities at Three Mile Island, demonstrates to the press corps the proposed defueling operation for the crippled reactor. Tools with long handles were specially crafted for the task of reaching down into the depths of the reactor core to extract the estimated 150 tons of uranium fuel and core rubble that remained within the inactive reactor.

In these images, a plant worker prepares to perform defueling work in the Unit 2 reactor building. Defueling, or removal of the debris from of the Unit 2 reactor, was performed from a platform positioned above the open reactor. Tools with long handles were manipulated to cut and remove debris from the reactor core below. The debris was then loaded into canisters that were transferred to the adjacent fuel handling building through the fuel transfer tube that connected the two buildings. The canisters were placed into storage racks and ultimately loaded into a shipping cask placed onto a railroad flatcar that was bound for the Idaho National Engineering Laboratory. What began in October 1985 to remove 100 tons of reactor fuel and 50 tons of reactor debris was not completed until January 1990.

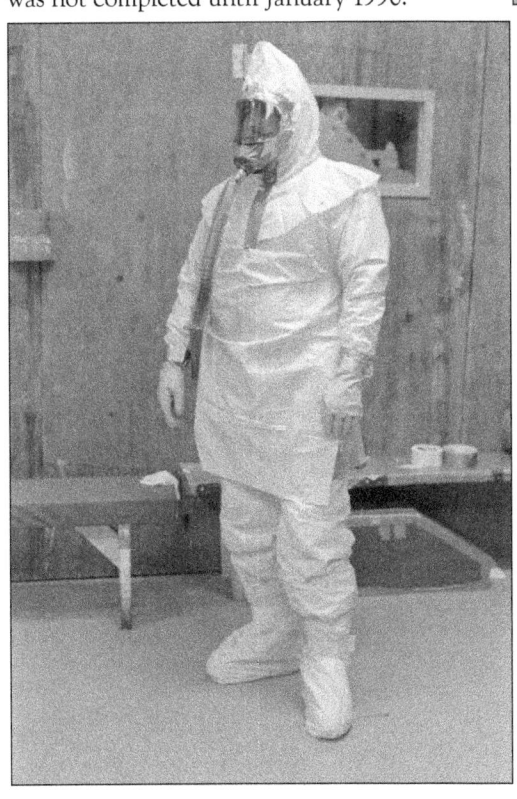

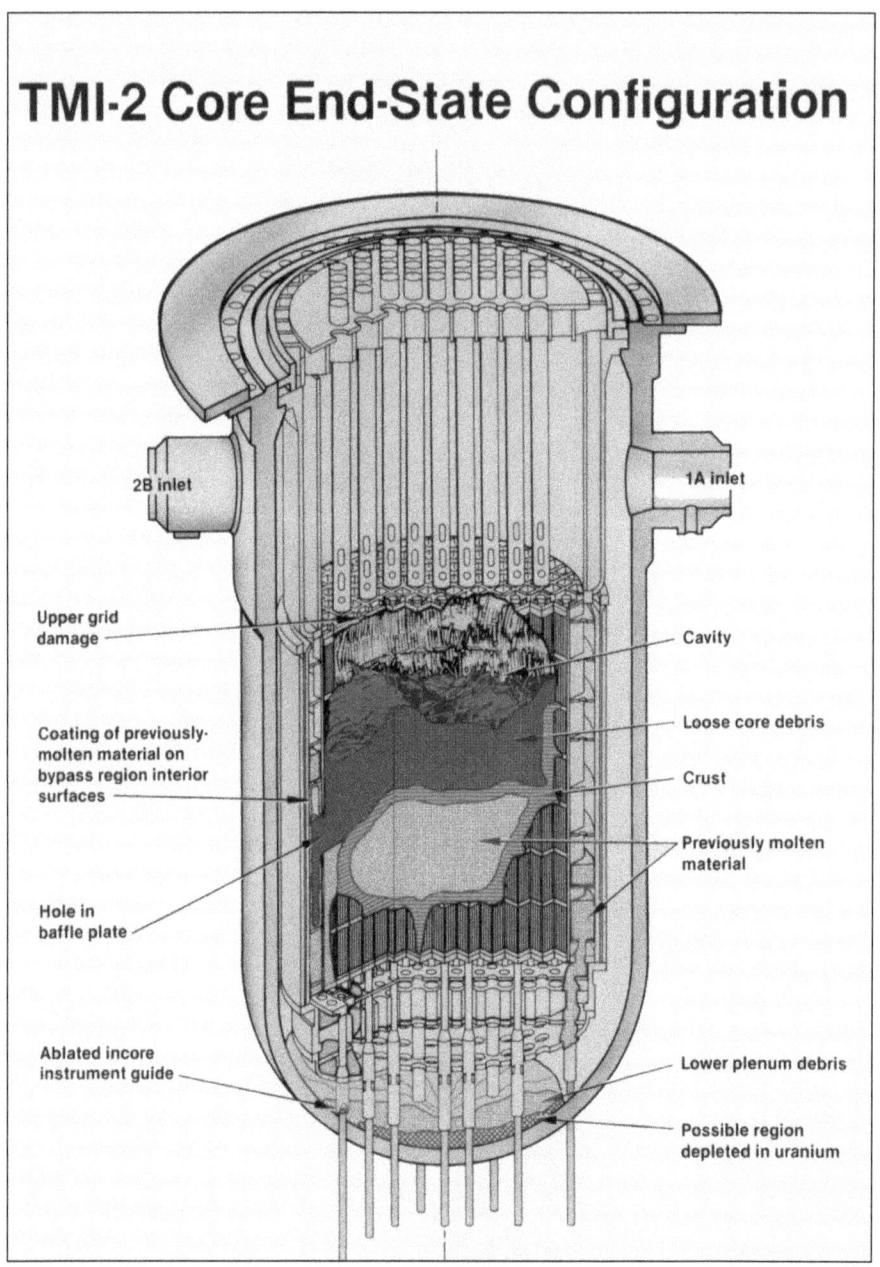

This diagram illustrates the damage sustained by the Unit 2 reactor core during the first hours of the accident that began on March 28, 1979. However, the extent of the damage was not known until after the reactor head was removed and cameras inspected the damage and defuelers began the cleanup operation. Only then was it confirmed that about 47 percent of the core material in the reactor had melted. Twenty-two percent of the core material had escaped the core boundary, with fifteen percent of the total core material escaping down to the lower plenum at the bottom of the reactor vessel. The cavity in the core was produced, in part, when portions of the core melted. It was also created when coolant was pumped back into the core three hours after the accident started and the "hot" core components shattered and created part of the bed of rubble later found during cleanup. (Courtesy of the Nuclear Regulatory Commission.)

Officials from the US Department of Energy and General Public Utilities held a press conference on Three Mile Island to show the stainless steel and lead shipping casks used to carry the canisters containing the damaged fuel and debris from the Unit 2 reactor. The dumbbell-shaped casks—shown, in part, on the flatbed railcar and in model form displayed in the foreground—are 23 feet long and 10 feet wide. The first shipment secretly left on July 20, 1986, and traveled from Pennsylvania through 10 states on its way to Idaho, and provisions were made to avoid major population centers such as Pittsburgh. Each trip, which would take four to five days, would transport a payload of approximately 10 tons of radioactive material. Below, the second shipment of radioactive material passes through Harrisburg on August 31, 1986.

Following a banner that reads, "A Decade of Delay, Deceit, and Danger . . . Three Mile Island 1979–1989," 150 protesters, some wearing gas masks, march along Route 441 in Londonderry Township toward the gates of the power plant on March 27, 1989, on the eve of the 10th anniversary of the nuclear accident. Gordon Tomb, spokesman for General Public Utilities (GPU), the owners of Three Mile Island in 1989, said, "We're inclined to take the opportunity to note the lessons from the Unit 2 accident and remind ourselves of the importance of remembering those lessons." Eric Epstein, speaking on behalf of Three Mile Island Alert, responded, "While GPU and the nuclear industry would like to look forward and forget March 28, 1979, this community will live in the shadow of the accident for the next several generations."

Speaking to the crowd gathered outside the gates to Three Mile Island on March 27, 1989, Frances Skolnick of the Susquehanna Valley Alliance told the crowd, "What you see are the human guinea pigs for the cleanup of a meltdown in the 20th century." The cleanup of the accident and partial meltdown was initially estimated to have a price tag of $40 million, but after more than 14 years, the final total cost to clean up the damaged Unit 2 reactor was $1 billion. And although the cleanup of Unit 2 was completed, the reactor remained contaminated and was effectively placed into storage until it could ultimately be decommissioned at a later date.

Just before 4:00 a.m. on March 28, 1989, eight individuals stood with lighted candles outside of the gates of the Three Mile Island nuclear power plant as an equal number of plant security guards stood and watched on the 10th anniversary of the nuclear accident. "The accident is not over," said Shawn Downey of Shermansdale, "that's the most important part of this vigil."

This photograph taken on March 16, 1989, next to South York Street and along the banks of Fishing Creek in Goldsboro shows a small, fenced-in area containing a radiation monitoring station that is capable of transmitting instantaneous radiation data back to a control center for processing. The system was installed in the fall of 1981.

On November 1, 1989, Joe Kuehn, Unit 2 site operations director, presented to members of the press a plan to begin the evaporation process for the 2.23 million gallons of radioactive water that had remained from the accident in March 1979. The process involved boiling the radioactive water to remove all contaminants except for tritium, which is a radioactive form of hydrogen. The tritium would be released, along with steam from the boiling process, out into the environment through a 100-foot-high venting stack. The gradual release of the tritium was estimated to provide a radiation dose that was comparably far less than that received from naturally occurring background radiation. The entire evaporation process was completed in August 1993 and carried a cost a $2 million.

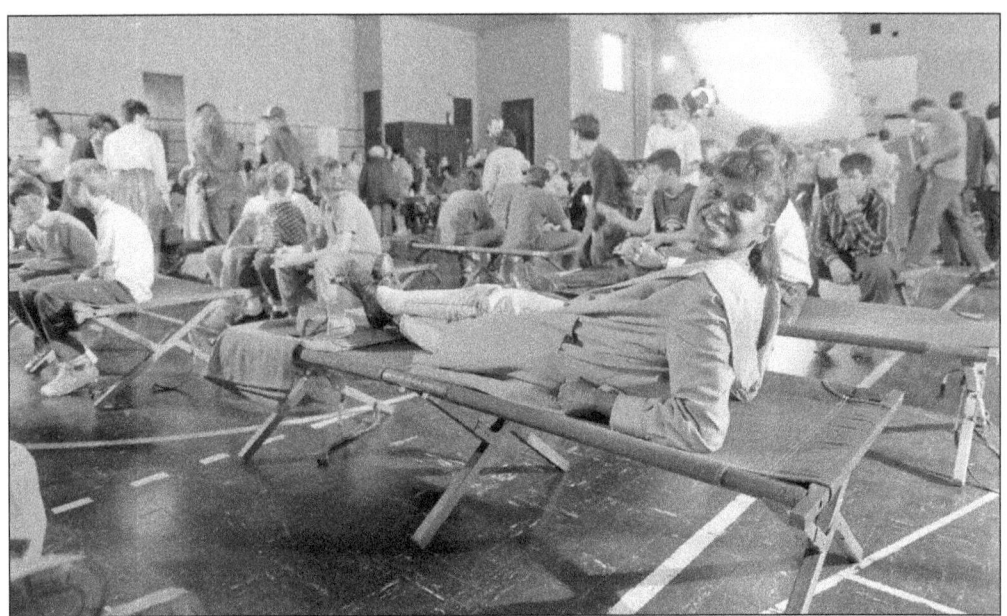

Although the nuclear accident at Three Mile Island had captivated the interest of the American public in 1979, the story was periodically revived for television network news programs. *Saturday Night with Connie Chung*, a program that aired on CBS, came to Middletown in October 1989 to film a re-created version of events that surrounded the nuclear accident. Above, the gymnasium in the Middletown Borough Hall building was one of the locations used to film scenes for the program. Below, a film crew sets up outside of the Three Mile Island observation center on September 25, 1992, to film scenes for the NBC program *What Happened?* Using data compiled by disaster research company Failure Analysis Associates, the program delved into what went wrong at the power plant and how it could have been avoided.

After two and a half years of slowly evaporating the 2.23 million gallons of radioactive water left over from the nuclear accident that occurred in March 1979, the cleanup process of the Unit 2 reactor was considered complete in August 1993. The total cost for the cleanup was estimated at $1 billion. This was the same as the cost to build the entire Unit 1 and Unit 2 nuclear power plants. Even after this sizable amount of money was spent to clean the damaged Unit 2, the containment building stayed sealed due to radioactive contamination levels that remained high within the building. Even though the radiation is sealed within the containment building, the specter of the power plant still hangs over the surrounding communities.

Visit us at
arcadiapublishing.com

www.ingramcontent.com/pod-product-compliance
Lightning Source LLC
Chambersburg PA
CBHW060938170426
43194CB00027B/2995